T0142253

Smart Innovation, Systems and Technologies

Volume 23

For further volumes:
http://www.springer.com/series/8767

Ioannis Hatzilygeroudis
Vasile Palade
Editors

Combinations of Intelligent Methods and Applications

Proceedings of the 3rd International Workshop, CIMA 2012, Montpellier, France, August 2012

Springer

Editors
Ioannis Hatzilygeroudis
Department of Computer Engineering
and Informatics, School of Engineering
University of Patras
Patras
Greece

Vasile Palade
Department of Computer Science
University of Oxford
Oxford
UK

ISSN 2190-3018 ISSN 2190-3026 (electronic)
ISBN 978-3-642-43503-4 ISBN 978-3-642-36651-2 (eBook)
DOI 10.1007/978-3-642-36651-2
Springer Heidelberg New York Dordrecht London

Printed on acid-free paper

Springer is part of Springer Science+Business Media (www.springer.com)

Preface

The combination of different intelligent methods is a very active research area in Artificial Intelligence (AI). The aim is to create integrated or hybrid methods that benefit from each of their components. It is generally believed that complex problems can be easily solved with such integrated or hybrid methods.

Some of the existing efforts combine what are called soft computing methods (fuzzy logic, neural networks, and genetic algorithms) either among themselves or with more traditional AI methods such as logic and rules. Another stream of efforts integrates case-based reasoning or machine learning with soft computing or traditional AI methods. Yet, another integrates agent-based approaches with logic and also nonsymbolic approaches. Some of the combinations have been quite important and more extensively used, like neuro-symbolic methods, neuro-fuzzy methods, and methods combining rule-based and case-based reasoning. However, there are other combinations that are still under investigation, such as those related to the Semantic Web. In some cases, combinations are based on first principles, whereas in other cases they are created in the context of specific applications.

The 3rd Workshop on "Combinations of Intelligent Methods and Applications" (CIMA 2012) was intended to become a forum for exchanging experience and ideas among researchers and practitioners who are dealing with combining intelligent methods either based on first principles or in the context of specific applications.

Important issues of the Workshop were (but not limited to) the following:

- Case-based reasoning integrations
- Genetic algorithms integrations
- Combinations for the Semantic Web
- Combinations and Web intelligence
- Combinations and Web mining
- Fuzzy-evolutionary systems
- Hybrid deterministic and stochastic optimization methods
- Hybrid knowledge representation approaches/systems
- Hybrid and distributed ontologies
- Information fusion techniques for hybrid intelligent systems
- Integrations of neural networks

- Intelligent agents Integrations
- Machine learning combinations
- Neuro-fuzzy approaches/systems
- Applications of combinations of intelligent methods to
  - Biology and bioinformatics
  - Education and distance learning
  - Medicine and health care

CIMA 2012 was held in conjunction with the 22nd European Conference on Artificial Intelligence (ECAI 2012).

This volume includes revised versions of the papers presented in CIMA 2012.

We would like to express our appreciation to all authors of submitted papers as well as to the members of CIMA-12 program committee for their excellent work. We would also like to thank the ECAI 2012 Workshop Chairs for accepting to host CIMA 2012.

We hope that this proceedings will be useful to both researchers and developers. Given the success of the first three Workshops on combinations of intelligent methods, we intend to continue our effort in the coming years.

Ioannis Hatzilygeroudis
Vasile Palade

Workshop Organization

Chairs-Organizers
Ioannis Hatzilygeroudis, University of Patras, Greece
Vasile Palade, University of Oxford, UK

Program Committee
Salha Alzahrani, Taif University, Saudi Arabia
Plamen Agelov, Lancaster University, UK
Soumya Banerjee, Birla Institute of Technology, India
Nick Bassiliades, Aristotle University of Thessaloniki, Greece
Kit Yan Chan, Curtin University of Technology, Australia
Artur S. d'Avila Garcez, City University, UK
Constantinos Koutsojannis, TEI of Patras, Greece
George Magoulas, Birkbeck College, University of London, UK
Toni Moreno, University Rovira i Virgili, Spain
Ciprian-Daniel Neagu, University of Bradford, UK
Jim Prentzas, Democritus University of Thrace, Greece
Roozbeh Razavi-Far, Universite Libre de Bruxelles, Belgium
Han Reichgelt, Southern Polytechnic State University, GA, USA
Jun Sun, Jiangnan University, China
George Tsihrintzis, University of Piraeus, Greece

Contact Chair
Ioannis Hatzilygeroudis
Department of Computer Engineering and Informatics
University of Patras, Greece
Email: ihatz@ceid.upatras.gr

Contents

Intelligent Agents: Integrating Multiple Components Through a Symbolic Structure

Razvan Dinu, Tiberiu Stratulat and Jacques Ferber

Abstract In order to handle complex situations, autonomous software agents need multiple components ranging from simple input/output modules to sophisticated AI techniques. Integrating a high number of heterogeneous components is a non-trivial task and this paper proposes the use of a symbolic middleware to handle inter-component interactions. A generalized hyper-graph model is defined, a simple and straightforward representation language is proposed and a pattern matching mechanism is introduced together with a basic performance evaluation. Finally, the paper shows how a flexible symbolic middleware can be built and a few examples are presented.

Keywords multi-agent system · multiple component integration · symbolic middleware

1 Introduction

In order to keep up with the increasingly complex real-world problems, autonomous software agents need to integrate more and more components that range from simple input/output modules to sophisticated AI techniques. As the number of components increases the integration itself becomes an issue which unfortunately has been neglected until recent years. More and more researchers agree that

---

R. Dinu (✉) · T. Stratulat · J. Ferber
Laboratoire d'Informatique, de Robotique et de Microelectronique de Montpellier, Montpellier, France
e-mail: dinu@lirmm.fr

T. Stratulat
e-mail: stratulat@lirmm.fr

J. Ferber
e-mail: ferber@lirmm.fr

I. Hatzilygeroudis and V. Palade (eds.), *Combinations of Intelligent Methods and Applications*, Smart Innovation, Systems and Technologies 23,
DOI: 10.1007/978-3-642-36651-2_1, © Springer-Verlag Berlin Heidelberg 2013

"the question about the inner workings of the pieces themselves holds equal importance to the question about the nature of the various dynamic glues that hold the pieces together" [16].

When integrating multiple components, two levels of integration can be distinguished: *generic* and *specific*. The generic level is concerned with general mechanisms such as *how* components communicate with each other and *how* they exchange data. The specific level is concerned with the details of integrating components X_1, $X_2, ..., X_n$, of specific types, such as *when* component X_i calls a function of component X_j, *what* data should X_i provide, *when* should X_i send the data to X_j, etc.

Usually, in a running software agent, the generic level takes the form of a middleware and provides primitives for data and control flow to different specific levels. Such a middleware has to provide solutions to three main challenges: *communication, data sharing* and *global control*. The communication challenge is concerned with how different components can reach each other and how can they use each other's functionalities. The data sharing is concerned with how components can provide data (content) to other components. Global control is concerned with how all the interactions between components are handled and how a coherent global behaviour of the agent can be achieved.

One traditional technique for generic integration of multiple components is the *blackboard system* in which a set of experts, also called knowledge sources (KS), are constantly monitoring a blackboard searching for an opportunity to apply their expertize. Whenever they find sufficient information on the blackboard they apply their contribution and the process continues [6]. Unlike other techniques that implement formal models, the blackboard approach was designed to deal with ill-defined complex interactions. One of the first applications of the blackboard system was the speech understanding HEARSAY-II system [7] in which multiple components used a shared blackboard to create the required data structures.

Another generic integration technique is based on message passing and usually uses a publish-subscribe mechanism in which components subscribe to different types of messages and whenever a message arrives it is forwarded to corresponding modules. A message-based communication protocol for AI that has been gaining in popularity in recent years is the OpenAIR protocol managed by mindmakers.org [2].

CORBA is a well known standard by OMG [5], according to which components written in multiple computer languages and running on multiple computers are exposed as objects and their interaction is performed by method invocation. CORBA is very used as system integration in humanoid robotics, see for instance the simulator OpenHRP [11].

All of the above techniques provide more or less solutions to the three challenges mentioned earlier. Blackboards clearly provide means for data sharing, enables communication between components indirectly, but the control component is usually a simple scheduler and it does not help much in assuring a coherent global behaviour of the agent. On the other hand, message-passing focuses on communication and object-oriented techniques on communication and somewhat data sharing. Both leave global control entirely up to the interacting components.

Both improvements and hybrid solutions have been proposed for the above techniques. For example, whiteboards [17] consist of a blackboard with (i) a general-purpose message type, (ii) ontologically defined message and data stream types and (iii) specification for routing between system components. They also add an explicit temporal model thus providing more specialized solutions for communication and data sharing challenges. Also, the GECA Framework (Generic Embodied Conversational Agent) uses a hybrid solution in which multiple blackboards are used to perform message-passing based on message types [12].

Our opinion is that components integration would be much easier if we had an integration technique based on a more expressive data model and which provided better support for different *patterns of global control.*

By pattern of global control we understand the most abstract model that can be used to explain the behaviour of the software agent. A classical example of such a pattern, especially used in robotics, is the Brooks subsumption architecture [4]. In this approach components are structured into layers and those situated at higher levels are capable of altering the input and inhibiting the output of components at lower levels.

Another very widely used pattern of global control, especially in multi-agent systems, is BDI (Beliefs Desires Intentions) [15]. The software agent maintains a set of beliefs based on which desires are created. A desire which the agent has decided to pursue becomes an intention and a plan is chosen to achieve the desired goal.

More sophisticated patterns of global control come from the agent architectures domain. For example the INTERRAP agent architecture [14] uses three control layers: Behaviour-based layer, Local Planning layer and Cooperative Planning layer. Each layer has its own world model and includes subcomponents for situation recognition, planning and scheduling.

When we say that the generic integration middleware should support patterns of global control such as the ones mentioned above we are not saying that the patterns should be entirely implemented inside the middleware. But rather, the middleware should contain only part of the pattern and should smoothly integrate with components implementing key aspects of the control pattern (for instance a planning engine).

This paper focuses on the generic level of integration and proposes a middleware model that enables easier and more straightforward integration of different AI and non-AI components of an autonomous software agent. The next section introduces our approach and Sects. 3–5 introduce our new symbolic model for generic integration and also perform a preliminary evaluation of its performance. Section 6 presents an implementation for smart phones based on the Android platform and finally, Sect. 6 and 7 present our comments and conclusions.

2 Approach

As it has been outlined in the previous section, the main shortcomings for current approaches concern the *data sharing* and *global control* challenges. Our approach is an extension of the blackboard model which addresses exactly these two challenges.

2.1 Data Sharing Challenge

Firstly, we propose that the blackboard uses a more expressive symbolic data model rather than just isolated bits of typed data. The chosen symbolic structure is inspired by the generalized hyper-graph model proposed by [3]. Hyper-graphs generalize normal graphs by allowing an edge to contain more than two nodes and a directed hyper-graph considers edges as ordered sets (tuples). We are interested in a generalization of directed hyper-graphs in which an edge can contain both nodes and other edges. This represents the generalized hyper-graph model we're using and it will be described in more details in the next section. However, we will be using a different terminology that makes more sense in the context of symbolic representations: *symbols* instead of nodes and *links* instead of hyper-edges.

Secondly, we extend the generalized hyper-graph structure with a *map* which associates each symbol of the hyper-graph with another symbol. Finally, we allow each symbol to have some attached information, which can be typed or not.

A generalized hyper-graph, the information associated with the symbols and the map of symbols form a SLiM structure (Symbol Link Map). From now on, we will use the capital version (SLiM) to refer to the model and the lower letter version (*slim*) as a shorthand for "SLiM structure" which refers to a concrete structure.

One related work which uses a hyper-graph model close to ours is [13]. They use a directed hyper-graph and integrate a typing system in which a node has a handle, a type, a value and a target set. The main differences in our model are the lack of the typing system and the addition of the symbolic map which, as it will be shown in future sections, can be used to create a typing system. However, they show how such a hyper-graph structure can be efficiently implemented and used as central database especially in AI applications. The OpenCog project [9] is also an illustrative example of hyper-graphs usage in AI projects. These works show the increasing interest of using the flexible hyper-graphs structures in AI.

2.2 Global Control Challenge

In order to address the issue of *global control* we inspired ourselves from the patternist philosophy of mind whose main premise is "the mind is made of patterns". In this perspective a mind is a collection of patterns associated with persistent dynamical processes capable of achieving different goals in an environment. For a quick overview of the patternist philosophy of mind and also a different way of applying it in the context of AI we recommend [8].

We define a pattern as a particular type of slim and we show how a set of patterns can be efficiently matched using an automaton. Next, we propose an interaction mechanism between components based on patterns that uses a central SLiM structure which can by accessed and modified by any component. Each component can register two types of patterns: data patterns and capability patterns.

Whenever a component modifies the central slim and a data pattern is found then the corresponding component is notified. Also, whenever a component requests the execution of something that matches a capability pattern then the corresponding component is notified.

As it will be detailed in the following section all these mechanisms provide a very flexible way of performing interaction between different components of a software agent and they can be packed into a symbolic middleware which can be used in conjunction with other agent frameworks.

3 The SLiM Model

This section formally introduces the SLiM model and also proposes a representation language to represent a slim.

3.1 Formal Definition

Definition 1 Let S be a finite set of elements. T_S is the set of all tuples over S and it is inductively defined as:

- $T_0 = \{(0, \emptyset)\}$.
- $T_k = \{X \cup \{(k, s)\} | X \in T_{k-1}, s \in S\}$ for $k \geq 1$
- $T_S = \cup_{k=0}^{\infty} T_k$

The sole element of T_0 is called the empty tuple and will be denoted simply by $\emptyset$. An element $t \in T_k$ is called a tuple of length k. Instead of $t = \{(0, \emptyset), (1, s_1), \ldots, (k, s_k)\}$ we use the equivalent notation $t = (s_1, s_2, \ldots, s_k)$. We also use the notation $s \in t$ to mean $\exists j \geq 1$ such that $(j, s) \in t$.

Definition 2 Let the following:

(i) S be a finite set of elements called *symbols*.
(ii) $l : S \rightarrow T_S$ be a function called a *linking* function on S.
(iii) $i : S \rightarrow I$ be a function called an *information* function on S, where I is a set of elements.
(iv) $m : S' \rightarrow S$, where $S' \subset S$, be a partial function on S called a *map* on S.

Then the quadruple $< S, l, i, m >$ is called a SLiM structure or simply a *slim*. The elements of $s \in S$ for which $l(s) \neq \emptyset$ are also called *links* and if $x, y \in S$ and $m(x) = y$ we say that x is mapped to y.

Below are a few terminological definitions associated with the SLiM model.

Definition 3 A symbol $d \in S$ is *reachable* from a symbol $s \in S$ if and only if there exist $s_1, s_2, ..., s_n \in S$ such that $s_1 \in l(s), s_2 \in l(s_1), ..., s_n \in l(s_{n-1})$ and $d \in l(s_n)$.

Definition 4 A symbol $d \in S$ is *mappable* from a symbol $s \in S$ if and only if d is equal to s or there exist $s_1, s_2, ..., s_n \in S$ such that $m(s) = s_1, m(s_1) = s_2, ..., m(s_{n-1}) = s_n$ and $m(s_n) = d$.

Definition 5 A tuple $(s_1, s_2, ..., s_n) \in T_S$ is called an *implied link* if and only if there exist $x_1, x_2, ..., x_n, y \in S$ such that x_1 is mappable from s_1, ..., x_n is mappable from s_n and $l(y) = (x_1, x_2, ..., x_n)$. If $x_i = s_i$ for $i = 1, n$ then the link is called *explicit*.

Definition 6 A slim $<S, l, i, m>$ is called *acyclic* if and only if no symbol can be reached from itself.

3.2 Representation Language

Before going any further we will introduce an abstract syntax for a representation language, called the *SLiM language*, that can be used to describe a slim. Given S and I the sets of symbols and information elements, the language is given by the following EBNF:

```
  slim → symbol+                              (1)
symbol → [id|link][:[info|symbol]]?           (2)
  link → [id=]?{symbol+}                      (3)
    id → s ∈ S                                (4)
  info → x ∈ I                                (5)
```

(the curly brackets are part of the terminal alphabet of the SLiM language)

Before giving a few examples we will provide the semantics of the production rules. In order to do that we consider that the non-terminal nodes of the grammar `symbol`, `link` and `id` have a synthesized attribute "s" which holds the corresponding symbol $s \in S$:

```
Production rule          Attribute rule
symbol → id[...]?        symbol.s = id.s
symbol → link[...]?      symbol.s = link.s
  link → id={symbol+}    link.s = id.s
  link → {symbol+}       link.s = use/new
    id → s ∈ S           id.s = s
```

The *use*/*new* keyword means that if there is already a link corresponding to the sequence of symbols on the right side then the attribute `link.s` uses the same id, otherwise it gets a random id from S not used by any other production rule. Below we will give the semantics of each of the right sides of production rules 2 and 3.

Right side	Semantics
{s1 ... sn}	$l(use/new) = (s_1.s, ..., s_n.s)$
id={s1 ... sn}	$l(id.s) = (s_1.s, ..., s_n.s)$
[id|link]:symbol	m([id|link].s) =symbol.s
[id|link]:info	i([id|link].s) =info

Here's an example of a slim described using the SLiM language:

```
here={my location}          (1)
here:{city Lyon}            (2)
{user said msg:"Hello"}     (3)
```

Let $<S, l, i, m>$ be the slim described in the above example. The symbols set is $S =$ { my, location, here, city, Lyon, user, said, msg, rand1, rand2 } and the information set is $I =$ { null, ''Hello'' }. The first line creates a link between the symbols my and location and assigns the id here, which means l(here) = (my , location). The second line creates a link between city and Lyon whose id is not important (and we can consider it to be rand1 $\in S$) and maps the symbol here to it. This means l(rand1) = (city , Lyon) and that m(here) = rand1. The third line creates a link between other three symbols and assigns some information to the last one, i(msg) = ''Hello''. All other symbols $s \in S$ have $i(s) =$ null.

The meaning of the first production rule is the union of all the symbols, information and mappings defined by the symbol production rules. We say that a SLiM representation (a string in the SLiM language) is *valid* if and only if there are no contradictions (i.e. a symbol being assigned two different information, a symbol being mapped to different symbols, multiple definitions of a link, etc.). From now on, we will use the SLiM language rather than the formal definition to describe SLiM structures.

One last observation concerns the use of curly brackets to create links. The language does not use parentheses or square brackets in order to avoid confusion with languages such as LISP or REBOL.

3.3 Modeling with SLiM

The SLiM model does not impose any particular semantics on the data being represented in a slim. It is a semi-structured, general purpose model based on a generalized hyper-graph structure. The actual schema that will be used in a slim will evolve dynamically and data integrity constraints can be enforced by different components using the slim. This type of flexible models is especially suitable for online environments.

4 Patterns

As it was discussed in Sect. 2, the pattern concept is central to our approach. Below we will explain what a pattern is, how can multiple patterns be matched, and we perform a simple performance evaluation of the proposed matching mechanism.

4.1 Definition

Definition 7 ? and @ are two special symbols called the *any* and the *root* symbols.

Definition 8 A *pattern* is an acyclic slim $<S, l, i, m>$ with the following properties:

(i) ? $\in S$ and @ $\in S$;
(ii) @ is mapped to a symbol, which is called the *root of the pattern*, and all other mappings, if any, are to ?;
(iii) the symbols mapped to ? are called *generic* symbols and they are all reachable from the root of the pattern;
(iv) it contains no information ($I = \{\emptyset\}$).

Before discussing the different properties from the above definition we will provide two examples, described using the SLiM language, so that the reader can have a better grasp of what patterns look like. Each pattern has been enclosed inside an additional set of curly brackets:

First of all, we are only interested in patterns at the symbolic level, disregarding the information attached to symbols, hence property iv). Secondly, the SLiM model is intended to represent data mainly through symbols and links and that is why patterns will be used to describe only parts of the symbolic hyper-graph. As a consequence, a pattern contains only mappings that have special meaning to the matching mechanism.

In a few words, a pattern is a generic way of describing a set of symbols and links. A pattern can contain regular symbols or *generic* symbols (those mapped to ?) which act as place holders for regular symbols. For convenience, we can omit the "@ : " for

Fig. 1 "Examples of patterns"

```
{
    {sound enabled}
    @: { notify user Message:? }
}
{
    online
    @: {User:? wants {listen album Album:?} }
    {User allowed music}
}
```

simple patterns. Also, we can write directly `{listen album ? }` if the name of the generic symbol is not relevant when presenting a pattern.

Intuitively the first example in Fig. 1 can be matched for example by the following slim:

```
 some-message: "Hello User!"
{sound enabled}
{notify user some-message}
```

The generic symbol `Message` is a place holder for the `some-message` symbol. The link `{sound enabled}` exists as it is.

In order to explain the role of the `@:` mapping we have to define exactly what is a match for a pattern.

4.2 Matching a Pattern

Definition 9 A slim M is a match for a pattern P if and only if it can be obtained from P by:

1. replacing all mappings to ? with mappings to other non-generic, possibly new, symbols;
2. replacing the occurrences of all generic symbols in links with the symbol they're mapped to;
3. removing the mapping from @ and the symbols ? and @.

Definition 10 Let X be a slim, P be a pattern and M be a match of P. We say that M is a match of P in X if and only:

(i) every symbol in M which is not generic in P also exists in X;
(ii) every link in M is implied in X.

For example, the following slim (M):

```
User: current-user
Album: s32
online
{current-user wants {listen album s32} }
{current-user allowed music}
```

is a match for the second example pattern in the following slim (X):

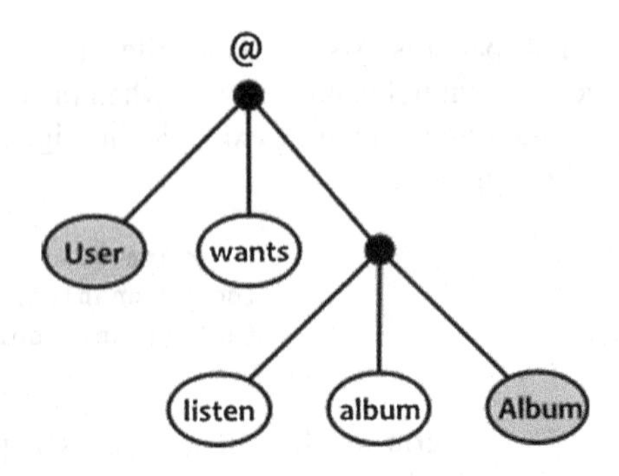

Fig. 2 Example of "tree of a pattern"

```
current-user: John
{s32 author Michael-Jackson}
{s32 title s41:"Bad"}
online
{current-user wants {listen album s32} }
{John allowed music}
```

We can easily see that the symbols such as `online`, `wants`, `music`, etc. exist directly in *X* and so do links such as `{listen album s32}` . On the other hand the link `{current-user allowed music}` is only implied in *X* (by definition 5) because `current-user` is mapped to `John` and we have the link `{John allowed music}`.

One important aspect of matching a pattern is the fact that the links described by the pattern must be implied in the slim we're searching, and not necessarily exist. This gives a lot of flexibility when modeling data we slim. We can choose to leave some links implicit and still be able to match them in patterns. A more in-depth discussion of implied links would be suitable but due to space limitation we will leave it to the reader to imagine how implied links can be used.

Searching a set of symbols and links that match a pattern in a slim can be a very time consuming task especially when generic symbols are used in more than one link. It is the equivalent of a join operation on a relational database which requires special indexing in order to be processed efficiently. That is why we divide a pattern in two, the root of the pattern and the rest, and we impose that the generic symbols are reachable from the root.

If the root is a link and some of the symbols inside the link are also links and so on, we have an ordered tree[1] of symbols determined by the root of the pattern. We will refer to this tree as *the tree of the pattern* (Fig. 2 shows the tree for the second example pattern).

Since all the generic symbols are reachable from the root, in order to find a match for a pattern we first have to find a match for the tree of the pattern. If one is found, all the other symbols and links can be easily checked since there are no

[1] a tree in which the order of the sons is important.

other generic symbols. But even finding a match for a tree in a slim can take a very long time so we will limit our approach to a particular case which will be used in our symbolic middleware presented in the next section.

The pattern matching problem we are addressing is the following: *given a symbol in a slim and a pattern tree, can it be matched starting at the given symbol (the root of the instance tree will correspond to the given symbol)*?

We will represent a pattern tree using the simplified notation (no `@` and no generic symbols names). For example, for the second example pattern we get `{? wants {listen album ?} }`. This will be called the normal string representation of a pattern tree. So, the question is, given such a representation and a symbol (which can be a link) in a slim, can the tree be matched with the given symbol as root? This can actually be done in liner time by reading the tree pattern representation from left to right and by performing a depth-first navigation of the symbol in parallel. The algorithm is left as an exercise to the reader.

4.3 Matching Against a Set of Patterns

Our approach of symbolic middleware will actually need to try to match a symbol against a set of pattern trees.

Given a set of pattern trees $P = \{p_1, p_2, ..., p_n\}$ and a symbol r we are interested in finding whether there is a pattern tree p_i such that it can be matched at r. We assume that the set of patterns P is independent, meaning that a match for one pattern is not a match for any other pattern.

One of the fastest way to perform tree matching is by creating a kind of deterministic matching automaton which is fed the nodes of the tree in a depth-first order for example. Graf [10] introduces an efficient technique to create a matching automaton based on the idea of *prefix unification*. This paper uses the same technique, which from now on will be referred to as TPMPU (Tree Pattern Matching based on Prefix Unification), slightly modified.

The main difference from the work in [10] is that according to definitions 10 we are only interested in matching the leaves of a pattern tree. The leaves correspond to symbols and they have to exist while the interior nodes (corresponding to links) can be matched by any symbol in a match (the definition only requires that they are implied which means they can exist with a different id or they can be implied by other links). Nodes corresponding to the `?` symbol in the pattern tree can be matched by any node in the match or even by a whole sub-tree. The `?` is the equivalent of the ω symbol in TPMPU.

Below we point out the required modification of the TPMPU technique (the next few paragraph require a good understanding of the TPMPU technique and are only given for readers familiar with it):

- instead of using $\Sigma - terms$ we use the normal string representation of a pattern in the SLiM language; also, let $\Sigma_S = S \cup \{ '\{', '\}', ? \}$ and Σ_S^* be the set of all normal string representations on the set of symbols S.
- the arity of a symbol s is pattern dependent and is implicitly given by the length of the tuple $l_p(s)$.

Also lemma 4.3 in [10] becomes:

Lemma 1 Let $\nu, \mu \in \Sigma_S^*$.

a $\nu \vee \mu = \top$ if either $\nu = \top$, $\mu = \top$ or ν and μ have different head symbols $\neq$?
b $\nu \vee \epsilon = \epsilon \vee \nu = \nu$
c $\alpha\nu \vee \alpha\mu = \alpha(\nu \vee \mu) \forall \alpha \in \Sigma_S$
d $? \; \nu \vee \alpha\mu = \alpha(\nu \vee \mu) \forall \alpha \in \Sigma_S \backslash \{?, '\{'\}$
e $? \; \nu \vee$ "{" $\mu =$ "{" $(?^{\sharp s}$"}"$\nu \vee \mu)$

where s is the link that starts at the "{" preceding μ, $\sharp s$ is its arity and $?^{\sharp s}$ is the ? symbol repeated $\sharp s$ times.

Also, a few other modifications are required such as in lemma 4.6 or 4.8 but they are analogous and will not be presented here due to space limitations. So, to summarize, we're using a different string representation for trees, which ignores the symbols corresponding to interior nodes, and modify the prefix unification algorithm accordingly. The rest of the technique (the way the closure is constructed and then the automaton) remains the same. Also, [10] shows how the closure for the set of patterns can be updated incrementally when new patters are added or existing patterns removed.

4.4 Time Complexity

After the automaton is created the complexity of the matching itself is independent of the number of tree patterns. The complexity of creating the automaton is mainly given by the size of the closure of P which itself depends on how much the tree pattern prefixes overlap. For a detailed complexity analysis [10] should be consulted which also gives some bounds on the size of the closure of P.

In the scenarios that will be presented in the next section the number of patterns that have overlapping prefixes is very limited which makes the above approach very efficient. However, one other important concern is the size of the automaton, mainly the size of its transition table. Below (see Figs. 3 and 4) we present some experimental results that show the number of patterns $|P|$, the number of patterns after performing the closure $|\bar{P}|$, the number of symbols $|S|$, the number of states of the automaton, the brute size of its transition table and the optimized size which will be explained below.

One key observation after preliminary experimental results was that, on the real-world data, the set of symbols S can be partitioned into $S = S_1 \cup S_2 \cup ... \cup S_n$

Fig. 3 Experimental results with pattern matching

$\lvert P\rvert$	$\lvert \bar{P}\rvert$	$\lvert S\rvert$	States	Brute size	Optimized
10	12	24	62	5 kb	1 Kb
50	105	82	367	117 Kb	7 Kb
100	208	122	706	336 Kb	15 Kb
200	412	196	1389	1063 Kb	31 Kb
300	620	277	2096	2267 Kb	38 Kb
400	823	352	2782	3825 Kb	67 Kb
500	1031	418	3374	5509 Kb	84 Kb

Fig. 4 Examples of used patterns

```
{send file ? by email to ?}
{for ? in ? do ?}
{is ? linked to ?}
{speak ?}
```

such that for every state the set of symbols on which transitions are defined is included in only one S_k. An optimized version of the algorithm is based on this observation and maintains multiple smaller transition tables. Since there are a lot of states with very few transitions the memory optimisation is very efficient.

In terms of speed, experimental results showed that the automaton was able to match about 1200 patterns per second on a mobile device with 1Ghz processor.

4.5 Summary

In order to conclude this section we will summarize the main idea: we have a set of patterns and a symbol in a slim and we are interested to see if there's a pattern whose tree can be matched with its root at the given symbol; if so, then the rest of the pattern can be easily checked once we have the values of all the generic symbols in the pattern; if all other symbols and links exist then we have found a match for a pattern. Finding if a tree can be matched at a given symbol can be efficiently done using an automaton, similar to the TPMPU technique.

5 SLiM Based Integration Middleware

Now we will show how the symbolic SLiM module introduced in the previous section can be used in what we call *symbolic integration middleware.*

5.1 Symbolic Middleware

We consider a software agent as having two parts: a *crown* which contains many different components and a *trunk* which contains one or more middlewares. The only way components can interact is through a middleware situated in the trunk. The SLiM middleware, which is the proposed solution for components integration, is composed of:

- A *slim* which acts as shared blackboard. Every component is able to create new symbols, modify links, attach information to symbols, change mappings, etc.
- A *capabilities index* and a *triggers index*. They are two sets of patterns together with two automata capable of matching them.
- A *behaviour rules* set. A rule is an association between a trigger pattern and an *entry-slim* (a slim with a designated symbol called *entry point*).

Every component can register one or more capability or trigger patterns with the SLiM middleware. For example a text-to-speech component can register the capability pattern `{speak ? english}`. A natural language processing component can register the trigger pattern `{user said ?}`.

In order to illustrate the role of each type of pattern or rule we will explain the different modes in which the SLiM middleware can be used.

***tell* mode**. In this mode a component can access the shared slim and perform whatever changes it needs. For example in this mode the speech recognition module would add `{user said msg:''Hello world!''}`.

***do* mode**. In this mode a component asks the SLiM middleware to do something by providing an entry-slim. The middleware will try to match the given slim, starting from the entry point, by a capability pattern. If one is found then the component that registered the pattern will be notified and it will be provided the match. If no pattern is found and the entry point is a link then all the symbols in the link will be successively used as entry points, creating a sequence of *do*-s. For example a component can request `{{wait 5 seconds} {speak msg:''Hello you too!'' english}}` to be done.

***trigger* mode**. Whenever a symbol is created or a link made, the triggers automaton will try to match it to an existing trigger pattern. If one is found and it is associated with a component then the component will be notified. However if it is associated with an entry-slim, through a behaviour rule, a *do* will be requested on the entry-slim. For example the rule `{user said ?}` $\rightarrow$ `{log to history ?}` logs what the user said to a history.

***ask* mode**. This mode is not discussed in this paper but it allows a module to query a slim structure in the same manner as [13].

Other details such as how symbols from a trigger pattern match are used in the entry-slim of a behavior rule, details of the *do* mode or the role of mappings, which some readers might consider important, are not discussed in this paper.

Let's take now the two example patterns provided in previous section (Fig. 1). The first one could be registered by a component capable of producing some sounds, perhaps depending on the type of the message. However, the pattern also contains the link `{sound enabled}`. We can imagine a convention like `{sound enabled}` when components can produce sounds and `{sound disabled}` when they can't (the volume is set to 0). So, if a component wants to send a notification to the user it will request a *do* on the following slim `{notify user msg12: ''New email!''}`. At that point the capability index will try to match the given slim, starting from its entry point which is the whole link, and will find the example pattern 1 as being a match, the corresponding component will be notified and for example the user will hear a beep.

The second pattern is more complicated and it can correspond to a component capable of playing albums from internet for example. But this pattern will be registered as a trigger pattern and not a capability pattern. For example a speech recognition component combined with a natural language processing module can recognize that the user wants to listen to an album and it will create a link stating that fact. When the link is created the trigger automaton will analyze the link and if the user is online and is allowed to play music than a match is found and the player component gets notified. If no component would have registered the second example pattern than the link created by the natural language processing module would have no effect and maybe it will be removed by a component that deletes links unused for a certain amount of time.

These examples should give the reader an idea of how the interactions between components with different functions will happen by using the proposed SLiM middleware.

5.2 Relationship to Other Integration Paradigms

We will now briefly show how other integration paradigms can be simulated in a straightforward manner.

For a message based paradigm each component can have an associated symbol C_k and register the trigger pattern `{message ? for` C_k`}`. For instance sending a message to a scheduler can be done by creating the link `{message {start jobX} for Cron}`.

For a publish-subscribe paradigm based on message types or topics each component can register the desired patterns as trigger patterns, for example `{message {type video} {content ?} }` or `{topic access-denied ?}`.

For an object oriented paradigm a component can create symbols for its objects and store its properties in the slim using links like `{objName propName propValue}` and expose its methods through patterns like `{objName methodName ? ? ?}`.

5.3 BDI Global Control

We will now show how global control can be implemented in the SLiM middleware by taking as example the BDI model.

The belief set can be represented directly in the slim through symbols and links. For example we can store information about the hotels in Lyon like this:

```
h={hotel Princess}
{h address}:''76 Rue Fesquet''
{h price-per-night}:''80 euros''
```

Goals can be represented by symbols organized in an explicit hierarchy.

```
{goal g1={prepare trip ?}}
{goal g2={find transport to ?}}
{goal g3={find accommodation in ?}}
{subgoal g2 g1}
{subgoal g3 g1}
```

Each goal can have a plan attached to it, for example `{plan g3 {search hotels in ?} {show options ? } }`

What we need is a small generic component capable of instantiating a hierarchy of goals. This module will look for patters of the form `{new goal ?}` and will analyze the hierarchy of goals of the given goal, instantiate it, and then launch the plans for each goal.

Through a method like the one briefly presented above one can program the global behaviour of a software system only by focusing on explicitly creating the hierarchy of goals and their associated plans which themselves use only trigger and capability patterns.

6 Test Implementation

The described SLiM middleware has been implemented and tested on a mobile device using the Android platform [1] which uses a message-based integration mechanism.

The implemented application asks the user the name of a city and then searches a predefined list of hotels. By showing the user a few options, and by integrating a simple clustering algorithm, the application is able to learn progressively which part of the city the user is interested in.

We had to integrate components already existing in the Android platform such as the text-to-speech engine, speech recognition, internet browsing and map view. For each of them we have created a wrapper that exposed the capabilities of each component through appropriate patterns such as `{show on map ?}` which looked for a link `{? address}` (where ? is the same in both patterns) and then showed the address using the available map view.

The application behaved correctly and the components integration was very smooth.

7 Comments and Limitations

We believe that integration of many of components is the key to making software agents smarter and make humans think of them as autonomous agents with which they can interact. Different works in AI and other connected domains are situated at different levels of abstraction and an integration middleware has to be able to deal with it. Also, the data that can be shared between different components is very diverse and an integration middleware would have to use a data model capable of handling this diversity. We believe the SLiM middleware, through the use of a very expressive data model and a flexible pattern matching mechanism, is a first step towards that.

One important limitation of the SLiM middleware comes from the fact that each created link or symbol has to be processed by the triggers index. This basically creates a bottleneck which means that slim updates cannot be performed at very high rates (on the android platform the maximum rate, as tested, is at about 1200 updates per second). This makes it not suitable for components that need interactions between them at a high rate. In that case, additional middlewares capable of handling such interactions should be used in the trunk together with the SLiM middleware. Alternatively, pattern checking can be performed in parallel on multiple cores which would also give better performance.

One interesting aspect of SLiM which has not been mentioned earlier is the fact that the SLiM language can be used as a scripting language for the interaction of components. We can have a default module which implements patterns for the usual control constructs we find in a scripting language (i.e. `{if Cond:? then Action: ?}`, `{for X:? in List:? do Action: ?}`, etc.). Together with the patterns registered by different components we will actually end up with a kind of domain specific language whose primitives are dictated by the capabilities of the agent. The following could be a perfectly valid slim for an agent that can be execute whenever `{time-of-day evening}`:

```
{
    {check weather}
    {if {tomorrow rain} then
        {warn user msg: "Take your umbrela"}
    }
}
```

Having a symbolic middleware like SLiM implemented in multiple agents, even developed by third parties, would create a more solid base for an ACL (Agent Communication Language). We can have the constructs of the language translated into symbolic representations which would then be executed by an agent using the registered components at a given time.

We are currently working on a prototype implementation of the middleware as a web application accessible on a server through a RESTful API. This approach has two possible advantages. On one hand we can create agents which have their components distributed over a network or even internet. On the other hand we can create lightweight agents which run for example in a browser and which hold the symbolic middleware on a server which has more computing power and which can perform pattern matching much faster.

As it can be seen, having an explicit symbolic middleware in a software agent has many advantages and it opens very interesting perspectives.

8 Conclusions and Future Works

This paper proposes the integration of multiple components in a software agent through the use of a symbolic middleware based on the new SLiM model. Due to the expressivity of hyper-graphs and the flexibility of the proposed pattern matching mechanism this model is well suited for the integration of AI with non-AI components.

We provided a clear formal description of the SLiM model, a representation language that can be used to describe slims and a pattern matching mechanism. Based on them we proposed a model of a symbolic middleware which uses a slim at its core and two types of patterns, capability and trigger patterns.

The efficiency of the pattern matching algorithm and its small memory footprint show that the SLiM integration middleware is well suited for mobile platforms such as Android on which a test implementation was done.

The proposed approach combines the advantages of multiple generic integration techniques: (a) the flexibility of a shared blackboard with an expressive hyper-graph based data model; (b) a triggering mechanism based on patterns which generalizes the publisher-subscribe paradigm; (c) *do* mechanism based on patterns similar to method invocation; (d) straightforward integration through behaviour rules based on patterns.

Also, the use of an explicit symbolic middleware such as the SLiM middleware can lead to interesting application such as a component integration scripting language, easier implementation of agent communication languages and even the creation of lightweight or distributed agents.

We believe future works should focus on a set of best practices on how to represent data in a slim and how to distribute the control between components and SLiM middleware. Common agreement is crucial to integrating components developed by different parties.

References

1. Android. http://www.android.com (2010).
2. OpenAIR. http://www.mindmakers.org (2011).

3. Boley, H.: Directed recursive labelnode hypergraphs: a new representation-language. Artifi. Intell. **9**(1), 49–85 (1977)
4. Brooks, R.A.: Intelligence without representation. Artifi. Intell. **47**(1–3), 139–159 (1991)
5. Corba, H.: (2011).
6. Engelmore, R., Morgan, A. (eds.): Blackboard Systems. Addison-Wesley, London (1988)
7. Erman, L.D., Lesser, V.R.: The HEARSAY-II speech understanding system: integrating knowledge to resolve uncertainty. Comput. Surv. **12**, 213–253 (1980)
8. Goertzel, B.: Patterns, hypergraphs and embodied general intelligence, International Joint Conference on Neural Networks, pp. 451–458. (2006)
9. Goertzel, B., de Garis, H., Pennachin, C., Geisweiller, N., Araujo, S., Pitt, J., Chen, S., Lian, R., Jiang, M., Yang, Y., Huang, D.: OpenCogBot: achieving generally intelligent virtual agent control and humanoid robotics via cognitive synergy. ICAI, (2010)
10. Graf, A.: Left-to-right tree pattern matching. In Book, R., (ed.) Rewriting Techniques and Applications, Lecture Notes in Computer Science, vol. 488, pp. 323–334. Springer, Berlin (1991)
11. Hirukawa, H., Kanehiro, F., Kajita, S.: OpenHRP: Open architecture humanoid robotics platform. In: Jarvis, R., Zelinsky, A., (eds.) Robotics Research, Springer Tracts in Advanced Robotics, vol. 6, pp. 99–112. Springer, Berlin (2003)
12. Huang, H.-H., Cerekovic, A., Pandzic, I., Nakano, Y., Nishida, T.: Scripting human-agent interactions in a generic ECA framework. In Applications and Innovations in Intelligent Systems XV, pp. 103–115. Springer, London (2008)
13. Iordanov, B.: HyperGraphDB: a generalized graph database. First International Workshop on Graph, Database (2010)
14. Muller, J.: The agent architecture INTERRRAP. In The Design of Intelligent Agents, LNCS, vol. 1177, pp. 45–123. Springer, Berlin (1996)
15. Rao, A.S., Georgeff, M.P.: BDI-agents: from theory to practice. In Proceedings of the First Intl. Conference on Multiagent Systems, San Francisco (1995)
16. Thórisson, K.: Integrated A.I. systems. Mind. Mach. **17**, 11–25 (2007)
17. Thórisson, K. R., List, T., Pennock, C., Dipirro, J.: Whiteboards: Scheduling blackboards for semantic routing of messages & streams. In AAAI-05 Workshop on Modular Construction of Human-Like, Intelligence, pp. 8–15, (2005)

An Architecture for Multi-Dimensional Temporal Abstraction Supporting Decision Making in Oil-Well Drilling

Odd Erik Gundersen and Frode Sørmo

Abstract We have developed an online decision support system that advice drilling engineers online based on data streams of real-time rig site measurements. This is achieved by combining multi-dimensional abstraction for recognizing symptoms and case based reasoning. Case-based reasoning compares the current situation in the well with past situations stored in the case base that contains advices for how to solve similar problems. The architecture is described in detail, and an example is presented in depth as well as results of commercial deployment.

Keywords case-based reasoning · real-data stream · oil-will drilling · integrated decision support

1 Introduction

Real-time data stream is available in a multitude of domains. Typical examples include health care in which patients are monitored continuously, stock market trading in which stock prices are recorded over time and aviation control. In this paper we will focus on the oil well drilling domain. Currently, over 3,800 oil-well drilling rigs are operating world-wide [11], and real-time data is collected from around 75 % of these according to domain experts. Examples of parameters that are measured in real-time are stand-pipe pressure, total amount of mud pumped through the drill-string, drill-string rotations per minute and the torque used by the motor to

O. E. Gundersen (✉) · F. Sørmo
Verdande Technology AS, 7041 Trondheim, Norway
e-mail: odderik@verdandetechnology.com
URL: http://www.verdandetechnology.com

O. E. Gundersen
Department of Computer and Information Science, Norwegian University of Science and Technology, Trondheim, Norway

I. Hatzilygeroudis and V. Palade (eds.), *Combinations of Intelligent Methods and Applications*, Smart Innovation, Systems and Technologies 23,
DOI: 10.1007/978-3-642-36651-2_2,

rotate the drill string. Although a vast amount of data is produced, there is still great unrealized potential in its use. Today, the data is primarily used for visual display, statistical analysis of performance and threshold alarming, and there exists a few systems that update physical models in real-time to get a better overview of the drilling environment [21], but these are not in wide-spread use. Monitoring is usually done manually by drilling engineers on the rig site, or in real-time operation centers that gather data from multiple rigs to a central location [5]. Typically, drilling engineers interpret these data continuously in shifts lasting up to 12 h by monitoring data graphs. Identifying situational events like symptoms of developing problems by manually inspecting data graphs is error prone, as it is a tiring task and it is hard to keep all previously identified symptoms in mind at all times.

Problematic oil-well drilling situations typically develop over time with symptoms that are recognizable by drilling experts. However, the symptoms behaves differently based on the context of the operation, like for instance with depth or the drilling tools used. Thus, threshold alarms, which are known to produce many false positive alarms, do not work well as users tend to ignore them if they go off often without any reason. Automatic analysis of the data from operations support drilling engineers by allowing them to shift the cognitive task from identifying situational elements to predicting the future status of the drilling situation and thus increasing the situational awareness helping drilling engineers to making better decisions [9].

Our main goal is to develop an online decision support system that provides relevant information to drilling engineers during oil well drilling operations using data streams of real-time rig site measurements and case-based reasoning. Case-based reasoning (CBR) is a methodology for reusing past experiences by comparing them to a current problem and then solves the current problem using the past experiences [2]. Experiences are stored in a case base as cases comprised of a problem description and a problem solution. A new experience is compared to the past cases in the case base. Then, the problem solutions of the most similar past cases are used to solve new experience. If no applicable cases are contained in the case base, the new experience can be retained along with the proper solution, which allows the system to learn by gradually building the case base.

In this paper we present an architecture that supports the main goal. The main component in the architecture is an agent system that coordinates a set of agents in which each agent has a specific task related to enhance the understanding of the current situation. Tasks include recognizing different aspects of the current operation, such as which activity is performed and symptoms of problems, in addition to projecting the future based on the recognized symptoms. In order to recognize different aspects of the operation multi-dimensional temporal abstraction is performed while CBR is utilized to project the future state of the operation and provide support in selecting the most appropriate remedial actions. Temporal abstraction is a term used when transferring data stream over time from a low level quantitative form to a high level qualitative form incorporating domain knowledge and context information [20], and hence avoiding simplistic threshold alarms that do not take sound reasons for parameters to change into account.

The architecture is implemented in DrillEdge, a commercial decision support system that is used by several large oil companies. In DrillEdge, different aspects of the current situation of the oil well drilling operation are visualized to alleviate the cognitive load of the drilling engineers monitoring the situation. Past problematic oil-well drilling situations and experiences related to how the drilling problem could be avoided are captured as cases, which are brought to the attention of the user when the current situation becomes similar to a past situation.

The rest of this paper is structured as follows. Some related work will be investigated in Sect. 2. The architecture of the system will be presented in depth in Sect. 3. In Sect. 5 selected results are presented, and finally in Sect. 6 we will conclude with some final remarks and future work.

2 Related Work

Cately et al. discuss trends and challenges of multi-dimensional temporal abstractions and data mining of medical time series in [6]. While they restrict their discussion to medical time series, several of the trends apply to the domain of oil well drilling as well. They identify six trends. Firstly, they note that data analysis methods must support multi-dimensional, high frequency real-time data. Multi-dimensional correlations between parameters exist in the oil well drilling domain, but while down hole tools record high frequency data, these are not communicated to the surface in real-time and can therefore not be used when supporting decisions made by humans. Another trend is that medical data mining systems and temporal abstractions should be applied to real-world clinical data. This clearly applies to real-world oil well drilling data as it often contains faulty readings and not only some occasional data points, but for long periods of time as the measuring tools are in though conditions. Furthermore, they require null-hypothesis testing of current frameworks. This also applies for oil well drilling data, but as far as we know, no work has been done in relation to this. Then the authors argue that data mining methods should be used to explore complex patterns that might hide parameter correlations from the users, which is true in the oil domain too. Some work has been done, but much is left to do, as oil well drilling companies are secretive about their data. Data mining algorithms will push the knowledge bases towards synthesized knowledge bases, that is knowledge derived through a data mining processes and not domain experts. This will apply to the oil well drilling domain when data mining methods get more widespread, but currently most knowledge is derived from domain experts. Finally, they identify a need for automatically transferring knowledge between data mining and temporal abstraction mechanisms, which is what we seek to achieve with the combination of pattern matching and CBR.

In [19], Stacey et al. presents an architecture for multi-dimensional temporal abstraction and its application to support neonatal intensive care. The goal was to go beyond threshold alarms causing many false positive alarms by introducing domain knowledge in form of an ontology and rules. The architecture was based on previously performed research in business event monitoring and performance measurement [13]. However, while the previous work focused on correlating workflow events, the neonatal intensive care research performs temporal abstraction on low level quantitative data from patient data streams logging parameters like blood pressure and oxygen saturation.

The temporal abstraction is performed by Patient Agents that execute rules on the patient data streams and fire alerts (events) if the conditions of the rules apply. Rules can be made manually by domain experts like doctors or mined automatically by an Analytical Processor. Such rules are supported as Smart Alarms in DrillEdge. However, because of the complexity of symptoms, rules will typically generate too many false alarms and thus we use Graph and Symptom Agents that are based on heuristic mathematical models designed by pattern matching experts in cooperation with oil-well drilling experts.

The alerts that are fired if a rule applies are received by an active ontology which triggers an alarm to the users of the system. The ontology is a central knowledge base which stores the rules and agent responses and allows temporal abstraction across multiple patient's data. A similar learning of patterns across operations is found in DrillEdge by introducing new cases representing new situations and experiences in the case base. While the active ontology performs this cross-correlating automatically, DrillEdge requires an intervention by the user to identify when to capture a case. The roots of the ontological reasoning in DrillEdge is from the Creek [1], a knowledge intensive CBR system in which ontologies are an integral part. Compared to Creek, the ontological reasoning mechanisms have been simplified in DrillEdge and the ontology is primarily used for quality assuring the terminology used by the system.

Montani et al. present an architecture in which a CBR system configures and processes temporal abstractions produced from raw time-series data in [14, 17]. The problem description part of cases capture the context knowledge about the time series interpretation while the problem solution part configure temporal abstraction trends and states that are used to monitor the time-series data. So, the CBR system configures the temporal abstractions using context information in order to monitor and evaluate patients undergoing hemodialysis to provide medical personnel with an assessment of the situation. In comparison, DrillEdge provide drilling engineers with an assessment of the current situation of an oil-well drilling operation by relating the temporal abstractions (events) with the context.

Another domain with large amounts of time-series data is in the financial market. Barbosa and Belo present an architecture for currency trading in [4], which will forecast the direction of the currency value the next six hours, decide how much to trade for and whether to trade. The architecture has three parts: An agent system which forecasts the direction of several currencies, a CBR system that decides how much to trade for based on previous experiences and a rule-based system that decides whether to trade or not. Each agent in the agent-based system forecasts the value of one currency by using an ensemble of data mining algorithms. The data mining algorithms perform the temporal abstractions which is the input to the CBR system. Then based on how coherent the data mining algorithms was in the decision that the agent made, the CBR system decides how much to trade. Feedback on the accuracy of the forecasting is used provided to the CBR system to learn whether to trust the decision in similar situations later. Thus, the system is autonomous and can learn from its own experiences in contrast to DrillEdge which requires manual intervention to capture a new case. The architecture was later used to implement an agent task force for stock trading [3].

3 The Architecture

The architecture is designed with online decision support and modularity in mind. From a process view, the architecture has four layers, which are: (1) *Data acquisition*, (2) *Data interpretation*, (3) *Decision support* and (4) *Visualization*. These four layers are implemented by (1) *Input Data Sources*, (2) *Data Analysis Agents*, (3) *Case-Based Reasoning* and (4) *Symptoms and Case Radàr* GUI elements, as illustrated in part *a* of Fig. 1. In a decision support system, the reasons for providing a recommendation must be transparent for a user to trust it, in particular when the users are experts in their field. Case-based reasoning systems provide such insight because it forms its advice based on a small (typically 1–5) set of similar cases, which include context and descritpion of the previous situation in a form natural to an expert. This allows the expert to verify that these cases are indeed similar to the current situation, and use the experience from them. The system is modular in that the system can grow with both symptoms and experiences without having to change what is already learned by the system. New agents recognizing new symptoms can be easily added and so can new cases.

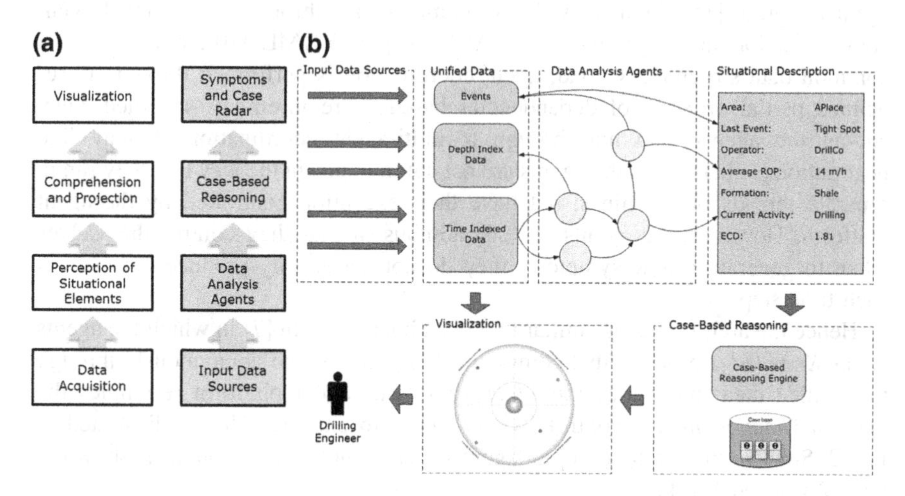

Fig. 1 **a** Layers in the architecture to the *left* and their implementations on the *right*. **b** Data flow oriented representation of the DrillEdge architecture

3.1 Data Flow

Part *b* of Fig. 1 shows how data flows through the system to the user. The data is gathered mainly from *Input Data Sources* such as streaming XML sources that use the oil well drilling industry standard called WITSML and the operators setting up and configuring operations that are to be monitored by DrillEdge. The different types of data that is stored in the system as *Unified Data* are time indexed data, depth indexed data and events, which are the multi-dimensional temporal abstraction that are computed by the *Data Analysis Agents*. In addition, the system stores contextual information describing the drilling operation. A structured *Situational Description* that describes the current state in the oil well is generated and compared by the *Case-Based Reasoning* engine to past cases with situational descriptions in the same structured format stored in the case base. The most similar cases are *Visualized* to the Drilling Engineer on the Case Radar as color coded dots according to their severity.

3.2 Agent Platform

The agent platform serves as an interface between the data and the agents in addition to control the execution of the agents. Agents depend both on the data streams and each other, and they are executed at every time step which is defined by the arrival of new data. Agents are executed in such a way that dependencies are maintained—i.e. an agent that depends on the output of another agent is always executed after the agent it depends on.

This agent architecture differs from the typical agent architecture found in the agent literature [15]. First, they do not communicate through a meta-level agent communication language like FIPA ACL [16] or KQML [10], but through a common data structure. Second, agent coordination is performed through a predefined partial ordering of dependencies based on requirements specified when coding the agents, and not through a multi-agent coordination strategy like negotiation [12]. Thirdly, the agents are not autonomous in the way that they run in separate threads or machines, but have their execution controlled by the agent platform. However, the agents are autonomous in that they control themselves when to generate a new symptom; they do not necessarily produce new values each time step.

Hence the architecture is similar to a blackboard system [7] in which the agents act as *Knowledge Sources* that communicate indirectly and anonymously through the unified data that act as the *Blackboard*. The agent platform resembles the *Control Component* that controls which agents to execute. This is illustrated in Fig. 2. Such an interpretation of an agent system is not uncommon, as is shown in Sect. 2 Related Work.

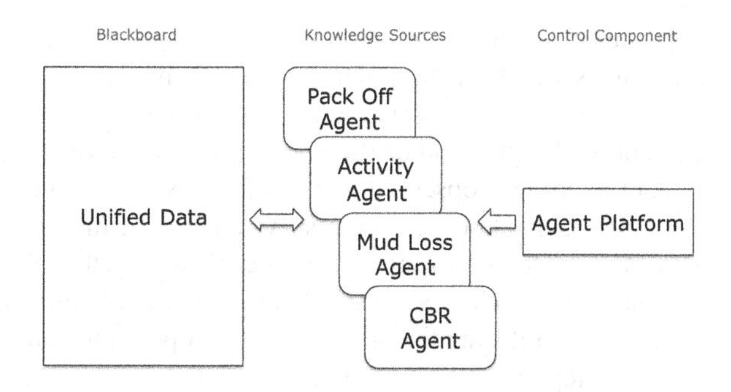

Fig. 2 The architecture can be interpreted as a blackboard system where the agents act as knowledge sources, the unified data can be thought of as the blackboard and the agent platform acts as the control component, controlling execution of the agents

We distinguish between three different types of data analysis agents in Drill-Edge -*Graph agents*, *Symptom agents* and the *CBR agent*. All types of agents can use all data available to the system to perform their tasks. Some agents only use parameter values stored in the time or depth indexes while others use context information too. The agent types and an example of dependence relations are illustrated in Fig. 3.

Graph Agents: In general, graph agents produce one or more new data value for every time step and store these in dedicated columns in the depth and/or time indexes. Graph agents are not necessarily dependent on any other agents, but some are. Mainly they produce values that are required by the other agents.

Symptom Agents: Symptom agents perform the multi-dimensional temporal abstraction by monitoring a set of predefined parameters and generate events if symptoms of problems are detected. The multi-dimensional temporal abstraction is implemented as fuzzy rules. An example of a symptom is when gravel starts packing off around the drill-bit. In the worst case, this can lead to the drill bit getting stuck in the ground. The symptom agent recognizing such pack-off tendencies perform multi dimensional temporal abstraction on several parameters, such as the amount of mud pumped in to the drill-string, standpipe pressure, torque and the weight on the drill-bit. In addition, it relies on context parameters like the dimension of the wellbore, casing depth and whether a mud motor is used when drilling. It is also dependent on the output from two graph agents, one that

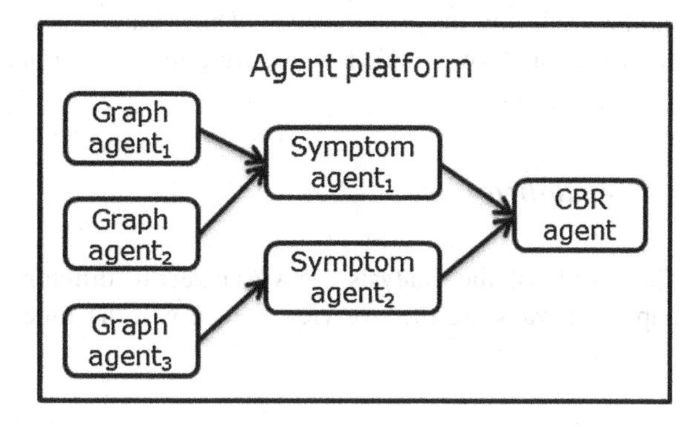

Fig. 3 The agent platform showing the three different agent types: graph agents, symptom agents and the CBR agent

identifies whether the drill-bit is in the open hole or in the casing and the agent that recognizes which drilling activity is performed.

When a symptom is recognized by a symptom agent, an event is produced, containing the time and depth where it occured, along with information about what type of event was observed and the severity of the event. Events serve as input to the CBR agent, but they are also visualized in the depth and time user interface view so that the end user can inspect them together with the parameters used to generate them. This is done to enhance the situation awareness of the drilling engineers monitoring the oil-well drilling operation and provides user transparency to the system.

CBR Agent: The CBR agent contains a CBR system and captures the current situation continuously and compares it to cases stored in the case base. The current situation is represented by a data structure containing both events and contextual information. Events are stored in the case as depth and time sequences sorted on the distance from the drill-bit and distance from the current time respectively. Problematic situations typically occur when symptoms cluster together—either on depth or in time or both. By storing the events both in time and depth sequences, the system can recognize when a case is similar along on or the other, using sequence similarity measures [18]. Context information describing the circumstances that the symptoms occurred in is also important. For instance, some problems only occur or is more likely to occur in particular geological formations.

3.3 Case-Based Reasoning Cycle

Figure 4 illustrates the CBR process in DrillEdge. Real-time data (1) is interpreted by graph and symptom agents and events are generated (2). Events together with other data are captured to form the input cases specifying the current situation (3). The case base is searched and past cases are retrieved from the case base (4). The past cases are plotted on the Case Radar to alert and advice the users (5). New cases can be captured and the retained case (6) is learned when stored in the case base.

Capturing and learning a new case is not done in real-time as quite a lot of experience is required to capture proper cases, and a new case should be tested on similar wells and the advice vetted. This means cases are typically captured after the operation is over, rather than during the monitoring.

3.4 Visualization

The results of the analysis are visualized in different ways, and the two most important ways are the overview screen and the time view. Users can verify or

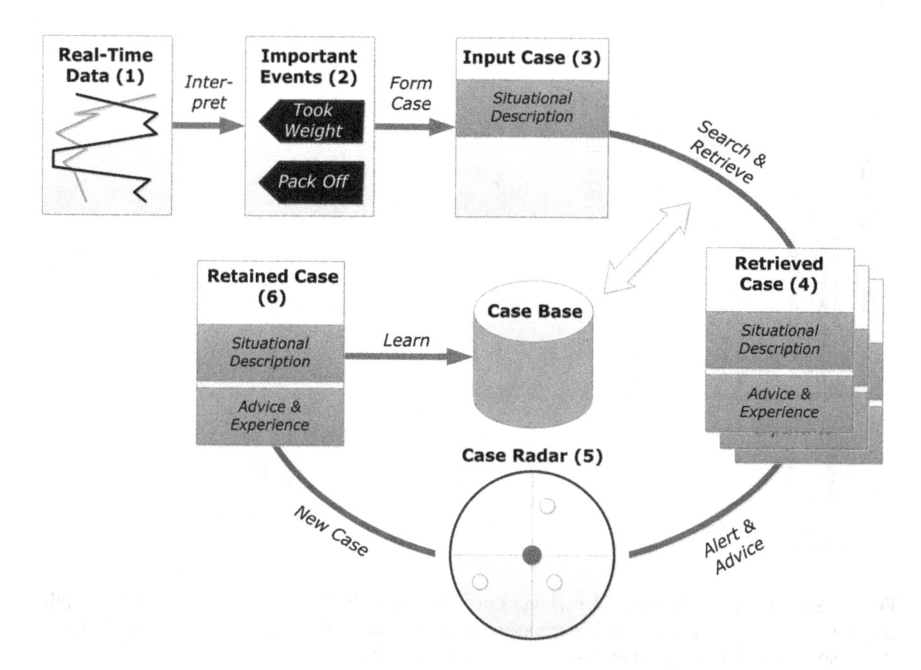

Fig. 4 Case-based reasoning cycle: from real-time data to a retained case

falsify the correctness of events by checking the parameter values around the time where the events are fired, and this is the main use for the time view. An screen capture of the time view is shown in Fig. 5.

Figure 6 shows a screen capture from the DrillEdge Client that visualizes the current state of an oil well drilling operation. The depth column with the drill-bit at the bottom of the wellbore and pack off events distributed along the drill string is

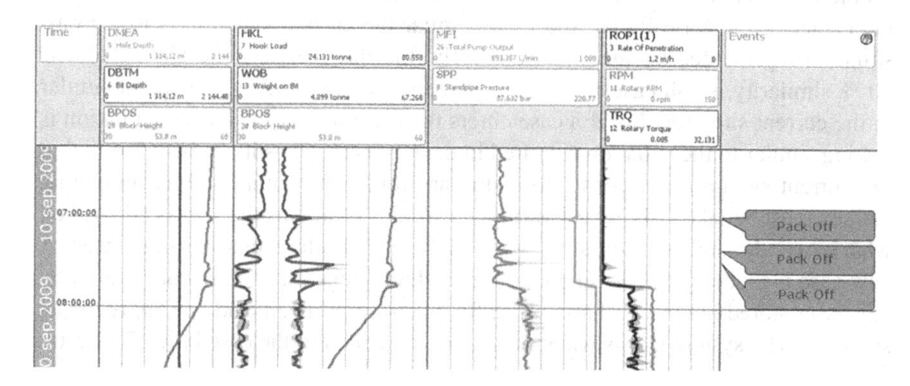

Fig. 5 Screen capture from the DrillEdge client showing the time view with parameters and events distributed over time

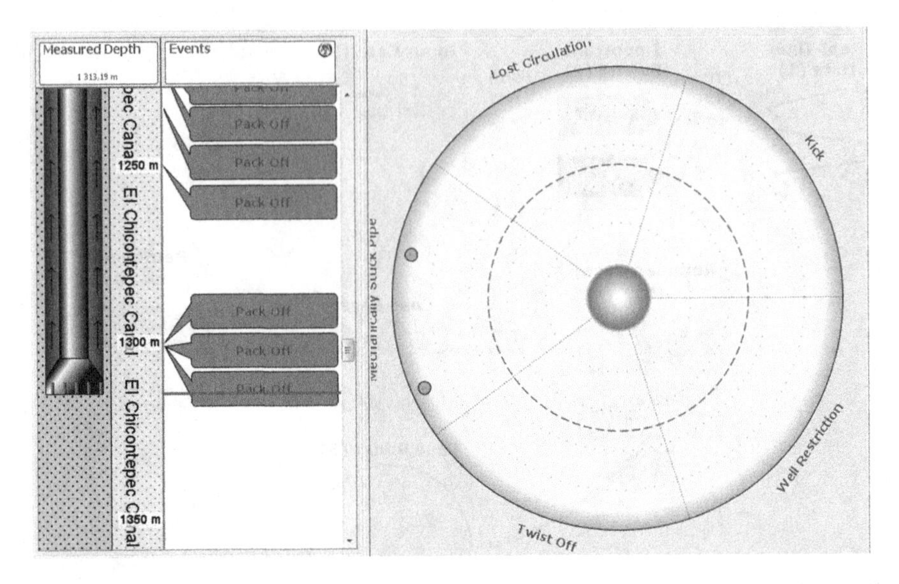

Fig. 6 Screen capture from the DrillEdge client showing the overview screen with the depth column containing events distributed according to depth and the Case Radar. Sections in the Case Radar specify problem type and case color indicate severity

showed to the left, and the Case Radar is depicted with two cases indicating a stuck pipe situation to the right.

The CBR agent searches for and retrieves cases from the case base and compares them to the current case. The comparison results in a similarity score for each case in the case base, describing how similar it is to the current case. All past cases with a degree of similarity above a given threshold are visualized on a GUI element, the Case Radar, to alert and advice the user of past historic cases that are similar to the current situation. The sectors in the Case Radar indicate different problem types that are covered by the cases in the case base. The closer a case is to the center of the Case Radar, the more similar the current situation is to the past situation represented by the case. Currently, the threshold for the Case Radar is 50 % similarity, so all cases on the Case Radar are more than fifty percent similar to the current situation. So, if a case enters the Case Radar, the current situation is getting similar to the situation captured in the past case. Thus it can be inferred that the current situation develops into a similar problematic situation and mitigating actions must be taken. By investigating the similar situations the user is advised on what others did and what should have been done in similar situations in the past. A new case is made by the drilling engineer if the current situation is not covered by any cases stored in the case base or if different or more specific advice apply to this situation. The system learns when new cases are added to the case base. The result of the case comparison is written by a graph agent to the time table so that the users can monitor the behavior of cases over time in the time view.

3.5 Client Server Architecture

The DrillEdge server is a distributed system that can be deployed on virtual server parks like the Amazon Elastic Computing Cloud (ECC) or VMWare clusters as well as physical servers. The client is a Java application that is installed from a web page using Java Web Start. All data interpretation is done on the server cluster, so the main task of the client is to visualize the information analyzed by the server.

The server is implemented as a set of services, and some services, like the license and management services, provide administrative functions and only one of each of these are required in the cluster. The license service ensures that the cluster runs with a valid license while the management service enables power users to set up and configure operations and administrators to administrate users. The operation service analyses oil-well drilling operations and communicates with DrillEdge clients through a front end. Each oil-well drilling operation monitored by the system has its own dedicated operation service.

Figure 7 illustrates the client server architecture. The box stippled with gray lines contain a server cluster deployed on the Amazon ECC while the gray boxes illustrate physical or virtual machines that run services. Hence, the server cluster is comprised of four machines in which two of them are running four operations, one machine is running the management and license server while the last one runs the front end service that communicates with the clients through HTTPS.

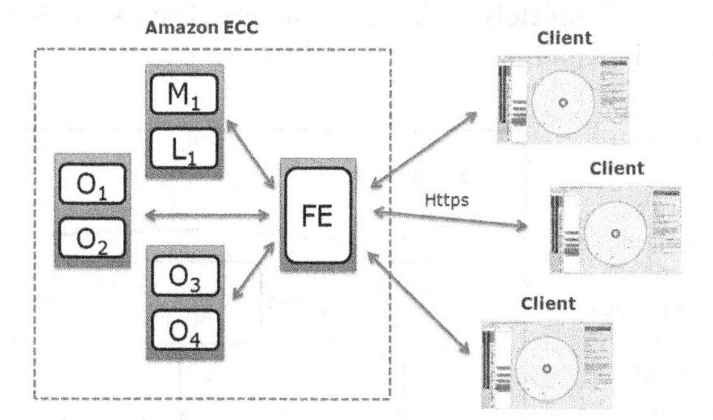

Fig. 7 The client server architecture: *O* indicates a operation service while *L* and *M* refers to license and management services respectively. *FE* is the front end and the subscripts indicate different services. The *gray*, stippled line confines the server cluster and the *gray boxes* indicate physical machines on the Amazon ECC

3.6 Operation Service

The main task of an operation service is to analyze the data streams from one concrete oil well drilling rig that has been assigned to the service. These data streams are measured on the rig, then aggregated and made available through a web service-based industry standard called WITSML [8]. The operation service is responsible for the data acquisition, and it has to communicate the results to the clients that users monitor the operations through. The agent platform is the most important part of the operation service.

Operation services are set up and configured by operators, which typically are drilling engineers responsible of ensuring that the service is set up correctly. The time streams update frequency ranges from a measurement per second to once per minute. However, the lower the data rate, the less detailed patterns can be recognized. DrillEdge achieves best results when the range between data points is shorter than ten seconds, as internal tests have shown that the results degrade significantly with lower frequencies. Depth data is measured by down hole tools, which collect measurements at high frequencies that are sampled and communicated to the surface at a rate of up to 40 bits per second. In addition to the streamed measurement data from the rig, the system requires context information that can be added to the operation service by operators. Context information includes static parameters like rig type and the configuration of the drilling tools, and the expected geology that is to be drilled through. When setting up the operation, the case base will have to be configured by the operators too.

4 Architecture and Decision Support Example

In this section, we will present an example showing how the architecture supports giving complex advices. Only three agents are required to illustrate how advices can be given for two completely different situations. The two different situations will be investigated in detail.

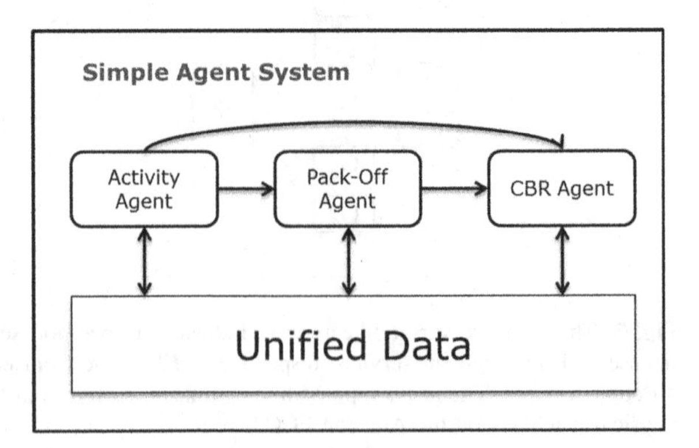

Fig. 8 The figure shows a simple agent system with one graph agent, the *Activity Agent*, one symptom agent, the *Pack Off Agent*, and the *CBR Agent*, which compares cases. Dependencies are illustrated by *arrows*. All the agents both read and write to the unified data

4.1 Agent System Configuration

The example system is illustrated in Fig. 8. It contains four components, and these are three agents that both read and write to the fourth component, unified data. Arrows indicate data flow, which also defines the dependencies. Three agents are implemented in the system: One graph agent, one symptom agent and the CBR agent. The four components of the system are described in detail below:

The Unified Data The unified data contains three types of data: Real-time parameters, symptoms, and context parameters. The following real-time parameters are used by the system:

- **Hole depth**: The depth of the hole that has been drilled. When this increase the hole is being drilled.
- **Bit depth**: Bit depth indicates where in the drilled hole the drill bit is located at a given point in time. If the bit depth is the same as the hole depth, the drill bit is at the bottom of the hole.
- **Rotation per minute**: Rotations per minute indicates whether the drillbit is rotating. If it is zero, then the drillbit is not rotating.
- **Mud Flow**: The amount of mud pumped into the well is indicated by mud flow.
- **Pressure**: Pressure indicates the pressure in the bottom of the well, and it depends on the amount of the mud pumped into the hole, which is measured by mud flow, and the length of the hole.

Only one symptom is recognized in the example system, which is pack off, and only one context parameter is provided, which is the formation type at a given depth.

The Activity Agent The Activity agent recognizes which activity that is performed on the rig. In this example, the activity code agent recognize two different types of activity. These are *drilling* and *tripping*. Drilling is when the drill bit is rotating at the bottom of the hole, while mud is flowing (to clean the cuttings of the drilling process out of the hole). Tripping is when the drill bit is moving into or out of the hole.

Algorithm 1 Activity Agent

$t \leftarrow getCurrentTime()$
$d \leftarrow getBitDepth(t)$
$d_1 \leftarrow getBitDepth(t-1)$
$h \leftarrow getHoleDepth(t)$
$m \leftarrow getMudFlow(t)$
$r \leftarrow getRotationPerMinute(t)$
if $d = h \wedge m > 0 \wedge r > 0$ **then return** isDrilling
end if
if $d \neq h \wedge d_1 - d \neq 0$ **then return** isTripping
end if

Algorithm 1 specify how drilling and tripping can be recognized. The current activity is recognized as drilling if the drill bit is at the bottom of the hole while mud is flowing and there is rotation. When the drill bit is not at the bottom of the hole, but is moving, the activity is recognized as tripping.

The Pack Off Agent Algorithm 2 lists a simplified pseudo code for the *PackOff* symptom agent, as presented in 3.2. The PackOff symptom agent generates a new pack off event at time t and depth d if the mudflow has been stable while the pressure has increased the last 60 seconds when drilling. The parameters $\boldsymbol{m}$ and $\boldsymbol{p}$ are vectors containing all values over the last 60 s of the mudflow and pressure parameters respectively. The methods *isStable*() and *isIncreasing*() can be implemented using regression analysis on moving windows of values that, for example, contains all values of a parameter over the last 60 s, such as $\boldsymbol{m}$ and $\boldsymbol{p}$.

Algorithm 2 PackOff Symptom Agent

$t \leftarrow getCurrentTime()$
$d \leftarrow getBitDepth(t)$
$t_0 \leftarrow t - 60s$
$\boldsymbol{m} \leftarrow mudflow(t_0, t)$
$\boldsymbol{p} \leftarrow pressure(t_0, t)$
if $isDrilling \wedge isStable(\boldsymbol{m}) \wedge isIncreasing(\boldsymbol{p})$ **then**
  return PackOff(t,d)
end if

The CBR Agent Data that is read by the CBR agent from the unified data are symptoms, activity and formation, which is a type of context information, while it writes the computed similarity for each of the cases in the case base. The CBR agent have a case base of two cases, which are illustrated in Fig. 9. *Case 1: Mechanically Stuck Pipe* captures a past situation in which several pack-offs were experienced and the outcome was that the drill string got stuck in the hole. The case description, which is used by the CBR agent to compare cases, contains a sequence of pack off events distributed over time and the activity, which was drilling at the time of the situation. When similar situations are experienced, the outcome might be the same. The action recommended by the experts capturing this case was to circulate—pump mud without drilling, to clean the cuttings out of the hole. The results of the case matching are written to the unified data in order to store the case matching history for later use.

Case 2: Severe Overpull describes a situation in which the drill bit was tripping into a section of the well that were not cleaned well enough when drilling. The case description contains the events distributed over depth close to the drill bit as well as the activity, which was tripping, and the formation, which was shale. When similar situations are experienced, tripping slowly is advice in order to avoid a severe overpull.

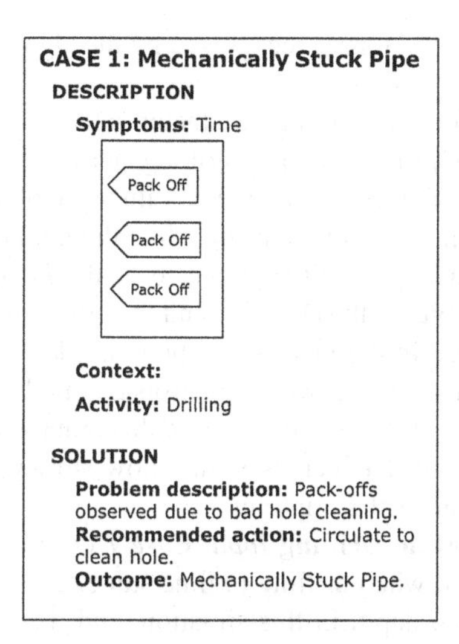

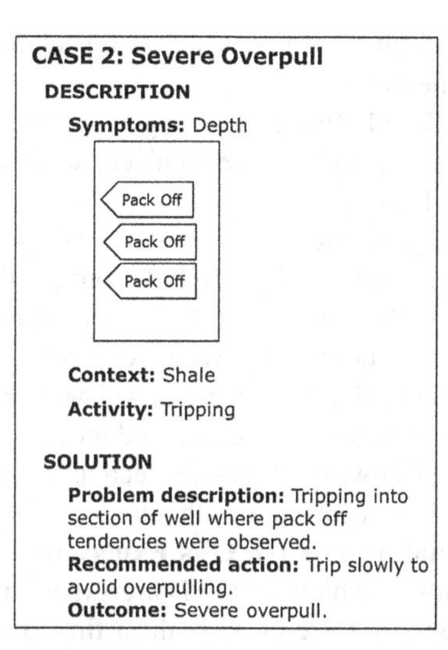

Fig. 9 *Case 1* describes a past situation in which the pipe got mechanically stuck while drilling due to cuttings packing around the drill bit, while *Case 2* describes a past situation in which a severe overpull was experienced when the drill bit was tripping through a section of the well where pack offs were observed when drilling. An overpull is a situation where more force than expected is required to pull the pipe out of whole, which can occur for instance if the drill bit get stuck on a ledge

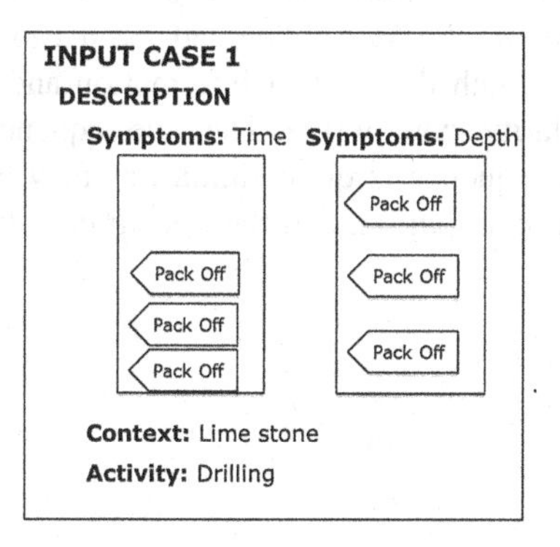

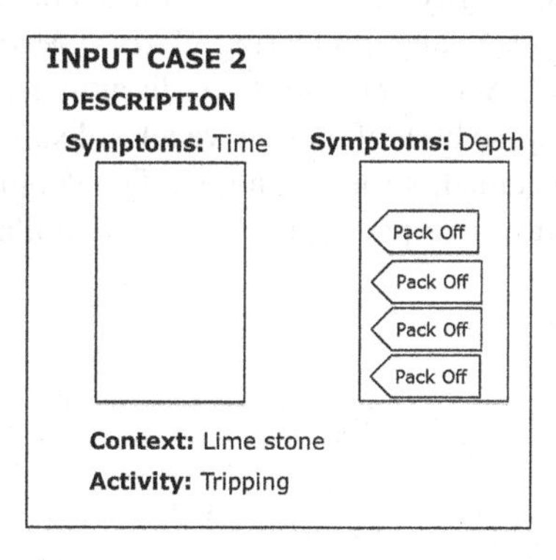

Fig. 10 *Input Case 1* describes a real-time situation in which pack offs are observed in lime stone while drilling, while *Input Case 2* describes another real-time situation in which the rig is tripping through a section of the well where pack offs were observed when drilling

4.2 Decision Support for Two Different Situations

The two input cases, which describe the two different situations that will be discussed in this section, are illustrated in Fig. 10. Each of the two situations will be analyzed by discussing how the two stored cases, which are retained in the case base of the CBR agent that was presented in 4.1, match in each situation. First, we will present an example of how a pack off symptom is recognized by the algorithm presented above.

Pack Off Recognition Fig. 11 shows some important parameter values around the time a pack off is recognized, so this is just a description of a generic pack off event. Time increases downwards on the y-axis while parameter values are plotted in a range on the x-axis. Pressure (SPP) can be seen increasing while the mud flow (MFI) is stable. This is right before the time a pack off is indicated by the *PackOff* agent. Hole depth (DMEA) is equal to the bit depth (DBTM), and thus the drill bit is at the bottom of the well. As can be seen in the graphs, the pressure decreases after the mud flow is decreased. The driller at the rig, which controls the mud flow, also recognizes the pack off and reduces the mud flow. Because of this reduction in the mud flow, the pressure decreases. The driller increases mud flow slowly and the drilling continues without any more pack off symptoms.

Situation 1: Pack Offs Experienced while Drilling *Input Case 1* describes a situation in which pack offs are experienced when drilling in lime stone. The pack off symptoms are close both in time and in depth in this situation, and there are three of them in the time sequence and three in the depth sequence, which is the same as for both of the stored cases illustrated in Fig. 9. The time sequence of *Case 1: Mechanically Stuck Pipe* is thus similar although not equal, as the distribution is somewhat different, and the similarity is computed to 85 %. The context and depth sequence are not specified in *Case 1*, so they are ignored, while the activity is drilling for both the input case and the stored case. Given that the total similarity is computed as the average of the features, the total similarity score is 92.5 % when comparing *Case 1: Mechanically Stuck Pipe* and *Input Case 1*. *Case 2: Severe Overpull* are dissimilar for both the context information and the activity, so both of these have a local similarity score of 0 %. The time sequence is not specified, so it is ignored. The depth sequence is quite similar, as they both contains three pack offs with quite similar distribution, and the system calculates

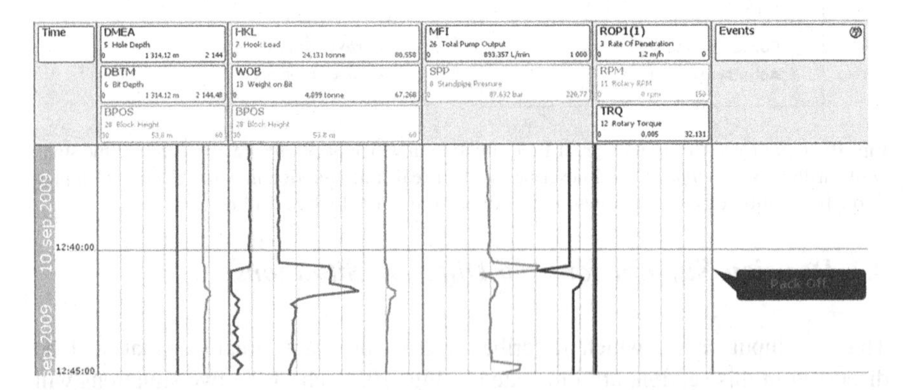

Fig. 11 A pack off event is recognized and shown in the time view when the pressure (SPP) increases while the mud flow (MFI) is stable

that there is a 90 % similarity. Even though the depth sequence is very similar, the total similarity score is just 30 %.

Therefore, in this situation the recommended action is to circulate to clean the hole in order to avoid getting mechanically stuck, as this is the recommendation from the most similar case, which is *Case 1: Mechanically Stuck Pipe*.

Situation 2: Tripping into a Section of the Well where Pack Offs Have Been Experienced Sitation 2 is specified by *Input Case 2*, which describe a situation where they are tripping into a section of the well where pack offs were experienced when the section was drilled. In this situation, no pack offs have been experienced over a certain time period, so the time sequence is empty. However, four pack offs were observed when drilling this section of lime stone, so the depth sequence contains four pack offs.

Case 1: Mechanically Stuck Pipe has three events in the time sequence, but as this is empty for the input case, the similarity score is computed to 0 %. Not context is specified in the stored case, so this is ignored, and the activity is dissimilar, which results in a 0 % similarity. The total similarity score for *Case 1: Mechanically Stuck Pipe* is therefore 0 %.

Case 2: Severe Overpull on the other hand, is found quite similar even though the context is dissimilar and has a similarity score of 0 %. The time sequence is not specified in the stored case, so it is ignored in the matching, but the activity is the same, and thus have a similarity of 100 %. The depth sequence in the input case have four pack offs and the stored case have three, but the distribution is similar, so the similarity is computed to 72 %. Hence, the total similarity between *Input Case 2* and *Case 2: Severe Overpull* is 57.3 %. This means that the recommended action in the situation represented by *Input Case 2* is to trip slowly to avoid a severe overpull.

4.3 Discussion

The example shows the strength of the architecture even with a small set of agents. It shows that complex recommendations can be given based on a small set of input parameters, as the *Activity* and the *Pack Off* agents perform multi-dimensional temporal abstraction on these parameters. Their output is used by the CBR agent, which provides the overall comprehension and predicts the outcome of the situation. This division between recognizing symptoms and diagnosing the situation has severeal advantages.

First, as many problematic situations in oil well drilling develop over a long time period and also are related to depth, a time series analysis only will provide poor results. Furthermore, given the many parameters that influence real-world situations, the number of examples required to train a machine learning system is enormous. Combining this with the scarcity of problematic situations reveals the impracticality of following this track. Drilling engineers are used to analyze time series data manually, and can easily verify or falsify single events when looking at the parameters relevant for a given symptom. Hence, the system is not thought of as a black-box system that just provides a recommendation without justifying it. The users can also see which symptoms are found relevant when comparing the current situation with a past case, which give them the possibility to decide themselves whether they agree or not. This increases the confidence of the users and they will more easily trust the advices than if no justification was given. Also, the modularity of the system makes it easier to expand to new problem types, as it only requires new agents to be deployed and new cases in the case base, which means that it can learn not only new situations, but also new problem types.

5 Results

The described architecture has been deployed in a commercial production environment since 2011. We give a short description of the deployment and present a case story from the production environment, which illustrate how the system can support decisions and save millions of dollars for oil well drilling companies.

5.1 Deployment

DrillEdge monitored the first live well during the summer of 2008 as part of the conclusion of the research project it was developed as part of. During 2011 around 15 pilot projects were finished, partly by analyzing historical data and monitoring live wells. In December 2011, Petroleum Development Oman (PDO) started monitoring live wells using DrillEdge, and in early January 2012 DrillEdge was running on more than 30 commercial wells concurrently. The current maximum number of operations monitored concurrently is 46, and it was reached in mid October 2012.

5.2 Case Story

Shell has run several tests both on historical and live data and deployed DrillEdge commercially in mid 2011. Shell experienced problems of twisting off the drill-pipe while drilling, and requested a solution that could predict twist off problems in advance. Twisting off the drill-string is a costly problem that often requires side tracking in order to avoid the parts of the drill-string that is left in the hole. Verdande Technology analyzed the data and found that long periods of maxing out the torque while drilling wore out the drill-string so that it finally twisted off. A symptom agent was developed to recognize when the torque was maxed out, and several cases were captured from historical data.

The solution was tested in a blind test using test data from a US land well while five cases was build from Middle East data. A total of 31 Maxed Out Torque events were fired, which resulted in three of the five Middle East cases appeared on the Case Radar before the drill-string twisted off. The first case appeared on the Case Radar two days before the twist off, potentially giving Shell long time to react to the problem. Cases were even captured from the US land blind test data and used to analyze the Middle East twist offs. Similar results were achieved. Other twist off tests were performed with varying results, but overall the tests were good enough for Shell to deploy DrillEdge commercially. In addition to the twist off tests, the stuck pipe solution was also put under pressure, but predicted stuck pipe six hours in advance. A detailed summary of the tests can be found in [22]. DrillEdge has been documented to accelerate learning in Shell [23].

6 Conclusion and Future Work

We have presented an architecture for multi-temporal abstraction that supports decision making in oil well drilling. The architecture scales well and is shown to predict problematic situations hours in advance. For this, Verdande Technology was awarded the Meritous Award for Engineering Excellence for its DrillEdge software platform by E&P Magazine in March 2011.

We will research how data mining techniques can be applied to detect symptoms of problematic drilling situations so that we do not need to rely on manually designed heuristic mathematical models. In order to do this, we need to be able to track event accuracy without human interference. Currently, we are implementing a system for quality ensuring event recognition automatically. Also, we will investigate how to combine a rule-based system with the case-based reasoning system in order to let domain experts control the case matching to a higher degree. Another task is to move in a more knowledge intensive direction and use the capabilities of the ontology as a part of the reasoning process. Furthermore, we would like to investigate ways to automatically identify where to capture cases and thus avoiding time consuming manual work. Finally, we are investigating whether this architecture and approach for data analysis and decision support can be expanded to other domains such as financial services and health care.

References

1. Aamodt, A.: Knowledge-intensive case-based reasoning in CREEK. In: Funk, P., González-Calero, P. (eds.) ECCBR, Lecture Notes in Computer Science, vol. 3155, pp. 1–15. Springer (2004)
2. Aamodt, A., Plaza, P.: Case-based reasoning: foundational issues, methodological variations, and system approaches. AI Commun. **7**(1), 39–59 (1994)

3. Barbosa, R., Belo, O.: An agent task force for stock trading. In: Demazeau, Y., Pechoucek, M., Corchado, J.M., Pérez, J.B. (eds.) PAAMS, Advances in Intelligent and Soft Computing, vol. 88, pp. 287–297. Springer (2011)
4. Barbosa, R.P., Belo, O.: Autonomous forex trading agents. In: Perner, P. (ed.) ICDM, Lecture Notes in Computer Science, vol. 5077, pp. 389–403. Springer (2008)
5. Booth, J.: Real-time drilling operations centers: a history of functionality and organizational purpose—the second generation. SPE Drill. Compl. **26**(2), 295–302 (2011)
6. Catley, C., Stratti, H., McGregor, C.: Multi-dimensional temporal abstraction and data mining of medical time series data: trends and challenges. In: Engineering in Medicine and Biology Society, 2008. EMBS 2008. 30th Annual International Conference of the IEEE, pp. 4322–4325 (2008)
7. Corkill, D.D.: Collaborating software: blackboard and multi-agent systems and the future. In: Proceedings of the International Lisp Conference. New York (2003)
8. Deeks, N.T.H.: WITSML changing the face of real-time. In: Intelligent Energy Conference and Exhibition (2008)
9. Endsley, M.R.: Toward a theory of situation awareness in dynamic systems. Hum. Factors J. Hum. Factors Ergon. Soc. **37**, 32–64(33) (1995)
10. Finin, T., Fritzson, R., McKay, D., McEntire, R.: KQML as an agent communication language. In: Proceedings of the Third International Conference on Information and Knowledge Management. CIKM '94, pp. 456–463, ACM, New York, NY, USA (1994), http://doi.acm.org/10.1145/191246.191322
11. Hughes, B.: Baker Hughes investor relations: overview and FAQ. http://investor.shareholder.com/bhi/rig_counts/rc_index.cfm (2012)
12. Lesser, V.R.: Reflections on the nature of multi-agent coordination and its implications for an agent architecture. Auton. Agent. Multi-Agent Syst. **1**, 89–111 (1998)
13. McGregor, C., Schiefer, J.: A web-service based framework for analyzing and measuring business performance. Inf. Syst. E-Bus. Manage. **2**(1), 89–110 (2004)
14. Montani, S., Bottrighi, A., Leonardi, G., Portinale, L.: A CBR-based, closed-loop architecture for temporal abstractions configuration. Comput. Intell. **25**(3), 235–249 (2009)
15. Nwana, H.S.: Software agents: an overview. Knowl. Eng. Rev. **11**, 1–40 (1996)
16. O'Brien, P.D., Nicol, R.C.: FIPA—towards a standard for software agents. BT Technol. J. **16**, 51–59 (1998)
17. Portinale, L., Montani, S., Bottrighi, A., Leonardi, G., Juárez, J.M.: A case-based architecture for temporal abstraction configuration and processing. In: ICTAI IEEE Computer Society, pp. 667–676 (2006)
18. Richter, M.: Similarity. In: Perner, P. (ed.) Case-Based Reasoning on Images and Signals, Studies in Computational Intelligence, vol. 73, pp. 25–90. Springer (2008)
19. Stacey, M., McGregor, C., Tracy, M.: An architecture for multi-dimensional temporal abstraction and its application to support neonatal intensive care. In: Engineering in Medicine and Biology Society, 2007. EMBS 2007. 29th Annual International Conference of the IEEE, pp. 3752–3756 (2007)
20. Stein, A., Musen, M.A., Shahar, Y.: Knowledge acquisition for temporal abstraction. In: Proceedings of AMIA Annual Fall Symposium. CIKM '94, pp. 204–208. ACM, New York, NY, USA (1996)
21. Strøm, S., Balov, M.K., Kjørholt, H., Gaasø, R., Vefring, E., Rommetveit, R.: The future drilling scenario. In: Offshore Technology Conference. CIKM '94, pp. 204–208. ACM, New York, NY, USA (2008)
22. van Oort, E., Brady, K.: Case-based reasoning system predicts twist-off in Louisiana well based on mideast analog. World Oil (2011)
23. van Oort, E., Griffith, J., Shneider, B.: How to accelerate drilling learning curves. In: SPE/IADC Drilling Conference and Exhibition (2011)

A New Impulse Noise Filtering Algorithm Based on a Neuro-Fuzzy Network

Yueyang Li, Haichi Luo and Jun Sun

Abstract A neuro-fuzzy network approach to impulse noise filtering for gray scale images is presented. The network is constructed by combining four neuro-fuzzy filters with a postprocessor. Each neuro-fuzzy filter is a first order Sugeno type fuzzy inference system with 4-inputs and 1-output. The proposed impulse noise filter consists of two modes of operation, namely, training and testing (filtering). As demonstrated by the experimental results, the proposed filter not only has the ability of noise attenuation but also possesses desirable capability of details preservation. It significantly outperforms other conventional filters.

Keywords Neuro-fuzzy system · Image processing · Filtering · Impulse noise

1 Introduction

Images are often corrupted by noise during the acquisition or transmission process. So noise cancellation/filtering is an important task in image processing, especially when the final product is used for edge detection, image segmentation, and data compression.

Image signals are composed of flat regional parts and abrupt changing areas, such as edges, which carry important information in visual perception. In the case of corruption by impulse noise, nonlinear techniques seem to perform better than linear ones, which tend to blur the edges and degrade the lines, edges, and other fine image details. So a great majority of filtering methods for the removal of impulse noise from images are based on median filtering techniques. The *standard median filter* (SMF) [1] is a simple nonlinear operation that outputs a median value

Y. Li (✉) · H. Luo · J. Sun
Key Laboratory of Advanced Process Control for Light Industry, Jiangnan University, Wuxi, China
e-mail: jsyueyangli@gmail.com

I. Hatzilygeroudis and V. Palade (eds.), *Combinations of Intelligent Methods and Applications*, Smart Innovation, Systems and Technologies 23,
DOI: 10.1007/978-3-642-36651-2_3,

of the pixels in a predefined filtering window to replace the center pixel of the window. The *weighted median filter* (WMF) [2] and the *center weighted median filter* (CWMF) [3] are extensions of the median filter, which give more weight to the appropriate pixels within the filtering window. These filters are spatially invariant operators, so they inevitably distort the uncorrupted pixels in image while restoring the corrupted pixels.

In the case of impulse noise removal, the aim of optimal filtering is to design noise reduction algorithms that would affect only corrupted image pixels, whereas the undistorted image pixels should be invariant under the filtering operation. So a number of algorithms, have been proposed to combine the median filter with a decision mechanism which attempts to determine whether the center pixel of a detecting window is corrupted or not. The *edge-detecting median filter* (EDMF) [4] has been proposed to combine the median filter with a decision mechanism which attempts to determine whether the center pixel of a detecting window is corrupted or not. A median based switching scheme, called *multi-state median filter* (MSMF) [5] is developed to adaptively switch among a group of center weighted median filters that have different weights by using a simple thresholding logic. The *signal-dependent rand-ordered mean filter* (SDROMF) [6] is conditioned on a state variable defined as the output of a classifier that operates on the differences between the input pixel and the remaining rand-ordered pixels in a sliding window. The performances of these filtering methods depend on one or more tuning parameters. In the filtering experiments, it is hard to choose the optimal values for these parameters and so the values of these parameters are heuristically determined.

In addition to the conventional filters discussed above, a number of filtering methods based on neural networks and fuzzy systems have been proposed. Fuzzy systems are fundamentally well suited to model the uncertainty that occurs when both noise cancellation and detail preservation are required. The *fuzzy filter* (FF) [7] is presented which adopts a fuzzy logic approach for the enhancement of images corrupted by impulse noise. On the other hand, artificial neural networks have the ability to learn from examples. Therefore, neuro-fuzzy systems combining neural networks and fuzzy set theories can be employed as powerful tools for the removal of impulse noise from digital images [8, 9].

In this paper, we propose a neuro-fuzzy (NF) network approach to impulse noise filtering for gray scale images. The network is constructed by combining four NF filters with a postprocessor. Each NF filter is a first order Sugeno type fuzzy inference system [10] with 4-inputs and 1-output. Each NF filter evaluates a different relation between the median value of all the pixels in a predefined selecting data window and the three appropriate pixels in a filtering window. The proposed impulse noise filter consists of two modes of operation, namely, training and testing (filtering). During training, each NF filter is trained individually by using a simple artificial training image and the internal parameters of each NF filter are adaptively optimized by training. During testing, the outputs of the four NF filters are fed to the postprocessor, which generates the final output of the network. As demonstrated by the experimental results, the proposed filter not only has the ability of noise attenuation but also possesses desirable capability of details preservation. It is compared with several conventional impulse noise filters. Simulation results show that the proposed filter significantly outperforms other conventional filters.

2 Algorithm

2.1 The Neuro-Fuzzy Network

Figure 1 shows the structure of the proposed NF network. The network is constructed by combining four NF filters with a postprocessor. Each of the four NF filters is a first order Sugeno type fuzzy inference system with 4-inputs and 1-output.

As to each pixel of the input image, the four inputs of each NF filter, x_1, x_2, x_3, and x_4, are obtained by employing a corresponding selecting data block, respectively. The procedure can be described as follows:

1. In Fig. 2, p_2 is the current operating pixel. Three pixels, p_1, p_2, and p_3, are obtained in the 3×3 pixel filtering window centered around p_2. Four different pixel neighborhood topologies are employed which correspond to four selecting data blocks shown in Fig. 1.
2. The median value m is obtained from all pixels in a predefined selecting data window centered around p_2. The method to determine the size of the selecting data window will be discussed later.
3. The four inputs of each NF filter, x_1, x_2, x_3, and x_4, can be defined as follows:

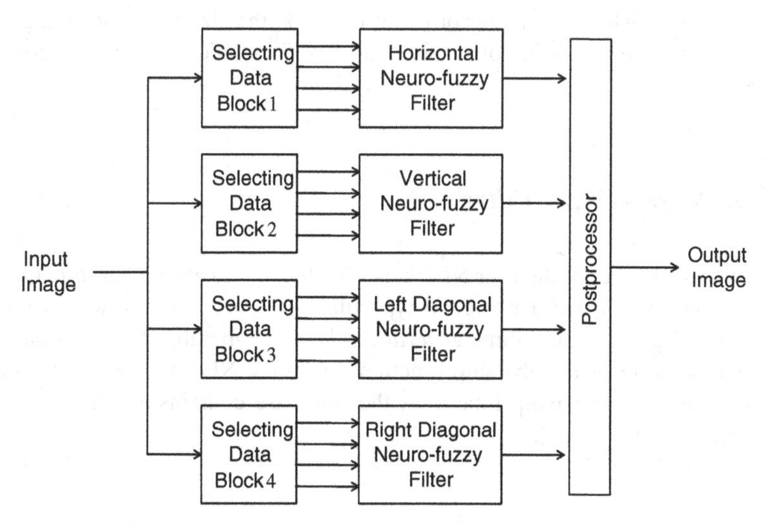

Fig. 1 Structure of the neuro-fuzzy network

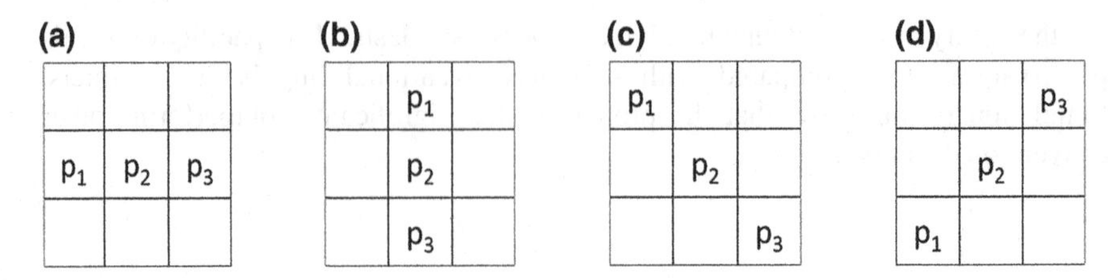

Fig. 2 Four pixel neighborhood topologies: **a** Horizontal direction; **b** Vertical direction; **c** Left diagonal direction; **d** Right diagonal direction

Table 1 The size of the selecting data window

Noise density (ND)	Size of the selecting data window
ND <= 10 %	3 × 3
10 % < ND < 40 %	5 × 5
ND >= 40 %	7 × 7

$$\begin{cases} x_1 = p_1 - m \\ x_2 = p_2 - m \\ x_3 = p_3 - m \\ x_4 = m \end{cases} \quad (1)$$

In order to obtain the median value *m*, the size of the selecting data window should be determined. Results of extensive simulation experiments indicate that the good filtering performance will be obtained if the size of the selecting data window is obtained in Table 1. According to Table 1, the size of the selecting data window is determined by the density of the noisy image. When the density of the noisy image is less than or equal to 10 %, the size of the selecting data window should be 3 × 3; when it is larger or equal to 40 %, the size of the selecting data window should be 7 × 7; otherwise, a 5 × 5 selecting data window is recommended.

2.2 The Neuro-Fuzzy Filter

The internal structures of the four NF filters are identical to each other. Each of the four NF filters is a first order Sugeno type fuzzy inference system with 4-inputs and 1-output. Each input has three *generalized bell* type membership functions and the output has a linear membership function. Since the NF filter has 4 inputs and each input has 3 membership functions, the rule base contains a total of 81 (3^4) rules, which are as follows:

Rule 1: if (x_1 is M_{11}) and (x_2 is M_{21}) and (x_3 is M_{31}) and (x_4 is M_{41})
then $y_1 = d_{11}x_1 + d_{12}x_2 + d_{13}x_3 + d_{14}x_4 + d_{15}$
Rule 2: if (x_1 is M_{11}) and (x_2 is M_{21}) and (x_3 is M_{31}) and (x_4 is M_{42})
then $y_2 = d_{21}x_1 + d_{22}x_2 + d_{23}x_3 + d_{24}x_4 + d_{25}$
Rule 3: if (x_1 is M_{11}) and (x_2 is M_{21}) and (x_3 is M_{31}) and (x_4 is M_{43})
then $y_3 = d_{31}x_1 + d_{32}x_2 + d_{33}x_3 + d_{34}x_4 + d_{35}$
Rule 4: if (x_1 is M_{11}) and (x_2 is M_{21}) and (x_3 is M_{32}) and (x_4 is M_{41})
then $y_4 = d_{41}x_1 + d_{42}x_2 + d_{43}x_3 + d_{44}x_4 + d_{45}$
Rule 5: if (x_1 is M_{11}) and (x_2 is M_{21}) and (x_3 is M_{32}) and (x_4 is M_{42})
then $y_5 = d_{51}x_1 + d_{52}x_2 + d_{53}x_3 + d_{54}x_4 + d_{55}$
Rule 6: if (x_1 is M_{11}) and (x_2 is M_{21}) and (x_3 is M_{32}) and (x_4 is M_{43})
then $y_6 = d_{61}x_1 + d_{62}x_2 + d_{63}x_3 + d_{64}x_4 + d_{65}$
⋮
Rule 81: if (x_1 is M_{13}) and (x_2 is M_{23}) and (x_3 is M_{33}) and (x_4 is M_{43})
then $y_{81} = d_{81,1}x_1 + d_{81,2}x_2 + d_{81,3}x_3 + d_{81,4}x_4 + d_{81,5}$

where M_{ij} denotes the jth membership function of the ith input, y_k denotes the output of the kth rule, d_{kl} is the consequent parameters, $i = 1,2,3,4$, $j = 1,2,3$, $k = 1,\cdots,81$, $l = 1,2,3,4,5$.

The input membership functions are generalized bell type:

$$M_{ij}(x_i) = \frac{1}{1 + \left|\frac{x_i - c_{ij}}{a_{ij}}\right|^{2b_{ij}}}, \tag{2}$$

where x_i is the ith input, a_{ij}, b_{ij}, and c_{ij}, are the antecedent parameters.

The output Y of each NF filter is the weighted average of the individual rule outputs y_k. The weighting factor w_k of each rule is the multiplication of four input membership values. Hence, the weighting factors $w_1,\cdots,w_{81}$ of the rules, and the output Y of each NF filter, can be calculated as follows:

$$\begin{aligned} w_1 &= M_{11}(x_1) \times M_{21}(x_2) \times M_{31}(x_3) \times M_{41}(x_4) \\ w_2 &= M_{11}(x_1) \times M_{21}(x_2) \times M_{31}(x_3) \times M_{42}(x_4) \\ w_3 &= M_{11}(x_1) \times M_{21}(x_2) \times M_{31}(x_3) \times M_{43}(x_4) \\ &\vdots \\ w_{81} &= M_{13}(x_1) \times M_{23}(x_2) \times M_{33}(x_3) \times M_{43}(x_4) \end{aligned} \tag{3}$$

$$Y = \frac{\sum_{k=1}^{81} w_k y_k}{\sum_{k=1}^{81} w_k} = \sum_{k=1}^{81} \bar{w}_k y_k \tag{4}$$

where $\overline{w}_k$ is the normalized weighting factor of the kth rule.

As shown in Fig. 1, the outputs of the four NF filters $Y_m(m = 1, 2, 3, 4)$ are fed to a postprocessor, which generates the final output of the NF network. The output of the postprocessor is calculated in two steps. First, the outputs of the four NF filters Y_m are truncated to the 8-bit integer values Y'_m so that the luminance values obtained range between 0 and 255.

$$Y'_m = \begin{cases} 0, & \text{if } Y_m > 0 \\ 255, & \text{if } Y_m < 255 \end{cases} \tag{5}$$

Then, the final output of the NF network Y_F is obtained by calculating the average value of the four NF filters.

$$Y_F = \text{round}\left(\frac{1}{4}\sum_{m=1}^{4} Y'_m\right) \tag{6}$$

where the function round(x) means to round the element x to the nearest integer.

2.3 Hybrid Learning Rule

The internal parameters of NF filter are optimized by using the hybrid learning rule [10] to reduce the error. The antecedent parameters (a_{ij}, b_{ij}, and c_{ij}) are optimized by using the gradient descent algorithm while the consequent parameters d_{kl} are solved by the least squares algorithm.

Assuming that the number of input–output training patterns is N. As to the current input $\left(x_1^t, x_2^t, x_3^t, x_4^t\right)^T$, Y^t is the output of the NF filter and Yd^t is the target output, $t = 1, 2, \cdots, N$. The superscript T denotes the matrix transpose. The objective error function can be defined as follows:

$$E = \sum_{t=1}^{N} E^t = \sum_{t=1}^{N}\left(\frac{1}{2}(Yd^t - Y^t)^2\right) \tag{7}$$

The gradient descent algorithm to minimize the objective error function mentioned above can be used to optimize the antecedent membership functions. The antecedent parameters (a_{ij}, b_{ij}, and c_{ij}) update rule are:

$$a_{ij}(n+1) = a_{ij}(n) - \alpha\frac{\partial E}{\partial a_{ij}} \tag{8}$$

$$b_{ij}(n+1) = b_{ij}(n) - \alpha\frac{\partial E}{\partial b_{ij}} \tag{9}$$

$$c_{ij}(n+1) = c_{ij}(n) - \alpha\frac{\partial E}{\partial c_{ij}} \tag{10}$$

where n denotes the learning epoch, α is the learning rate, $i = 1, 2, 3, 4, j = 1, 2, 3$. The initial value of the antecedent parameters (a_{ij}, b_{ij}, and c_{ij}) can be determined by the input–output training patterns, which will be discussed in detail in the following section.

The gradient descent algorithm can also be used to optimize the consequent membership functions. However, in order to improve the performance and speed of the optimization, here we use the least squares regression technique to solve for the consequent parameters d_{kl} rather than iteratively updating them by using the backpropagation algorithm. Because it uses two very different algorithms to reduce the error, the training rule is called a hybrid.

Let us rearrange the Eq. (4) into a more usable form:

$$\begin{aligned} Y = & \sum_{k=1}^{81} \bar{w}_k y_k \\ = & \bar{w}_1 y_1 + \bar{w}_2 y_2 + \cdots + \bar{w}_{81} y_{81} \\ = & \bar{w}_1 (d_{11} x_1 + d_{12} x_2 + d_{13} x_3 + d_{14} x_4 + d_{15}) + \\ & \bar{w}_2 (d_{21} x_1 + d_{22} x_2 + d_{23} x_3 + d_{24} x_4 + d_{25}) \\ & \cdots \\ & \bar{w}_{81} \left(d_{81,1} x_1 + d_{81,2} x_2 + d_{81,3} x_3 + d_{81,4} x_4 + d_{81,5} \right) \\ = & [\bar{w}_1 x_1 \; \bar{w}_1 x_2 \; \bar{w}_1 x_3 \; \bar{w}_1 x_4 \; \bar{w}_1 \; \bar{w}_2 x_1 \cdots \bar{w}_{81} x_1 \; \bar{w}_{81} x_2 \; \bar{w}_{81} x_3 \; \bar{w}_{81} x_4 \; \bar{w}_{81}] \\ & \times [d_{11} \quad d_{12} \quad d_{13} \quad d_{14} \quad d_{15} \quad \cdots \quad d_{81,1} \quad d_{81,2} \quad d_{81,3} \quad d_{81,4} \quad d_{81,5}]^T \\ = & \mathbf{XW} \end{aligned} \tag{11}$$

When input–output training patterns exist, the weight vector **W**, which consist of the consequent parameters d_{kl}, can be solved for using the least squares regression technique. So we can use the following equation to update the consequent parameters d_{kl}.

$$\mathbf{Yd} = \mathbf{XW} \tag{12}$$

where

$$\mathbf{Yd} = \begin{bmatrix} Yd^1 \\ Yd^2 \\ \vdots \\ Yd^N \end{bmatrix} \tag{13}$$

$$\mathbf{X} = \begin{bmatrix} \bar{w}_1^1 x_1^1 & \bar{w}_1^1 x_2^1 & \bar{w}_1^1 x_3^1 & \cdots & \bar{w}_{81}^1 x_4^1 & \bar{w}_{81}^1 \\ \vdots & & & & & \\ \bar{w}_1^N x_1^N & \bar{w}_1^N x_2^N & \bar{w}_1^N x_3^N & \cdots & \bar{w}_{81}^N x_4^N & \bar{w}_{81}^N \end{bmatrix} \tag{14}$$

$$\mathrm{W} = \begin{bmatrix} d_{11} \\ d_{12} \\ \vdots \\ d_{81,5} \end{bmatrix} \tag{15}$$

where Yd^t is the target output of the NF filter for the input $\left(x_1^t, x_2^t, x_3^t, x_4^t\right)^T$ and N is the number of input–output training patterns.

If the **X** matrix were invertible, it would be easy to solve for **W**:

$$\mathbf{W} = \mathbf{X}^{-1} \times \mathbf{Yd} \tag{16}$$

When the **X** matrix were not invertible, a pseudo-inverse may be used to solve for **W**. Better regression method uses the singular value decomposition (SVD) [11]. Here we use MATLAB pinv() function that automatically calculates a pseudo-inverse the SVD [12].

2.4 Training Procedure

The four NF filters shown in Fig. 1 operate on the same 3×3 filtering window and they are trained individually. Figure 3 shows the structure of training of an individual NF filter. The internal parameters of each NF filter are adaptively optimized by using a simple artificial training image. The training images of the four NF filters are same. The images shown in Fig. 4a and b are employed as the target and the input training images, respectively. The original image (target training image) is a 64×64 pixels image that can be easily generated in a computer [8]. Each square box in this image has a size of 4×4 pixels and the 16 pixels contained within each box have the same luminance value, which is an 8-bit integer number uniformly distributed between 0 and 255. The impulse noise image (input training image) is obtained by corrupting the original image by impulse noise of 30 % noise density.

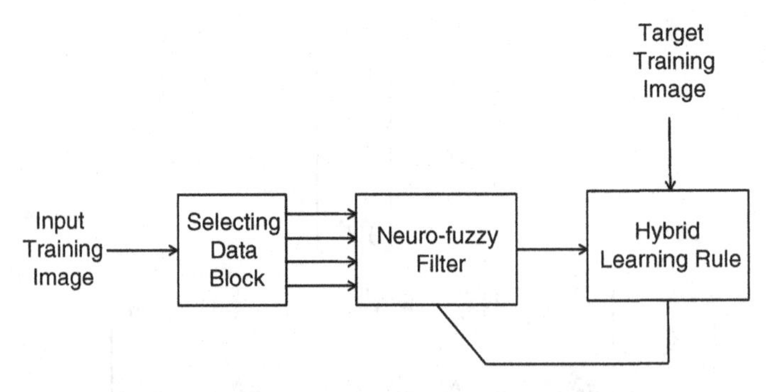

Fig. 3 Structure of training of an individual neuro-fuzzy filter

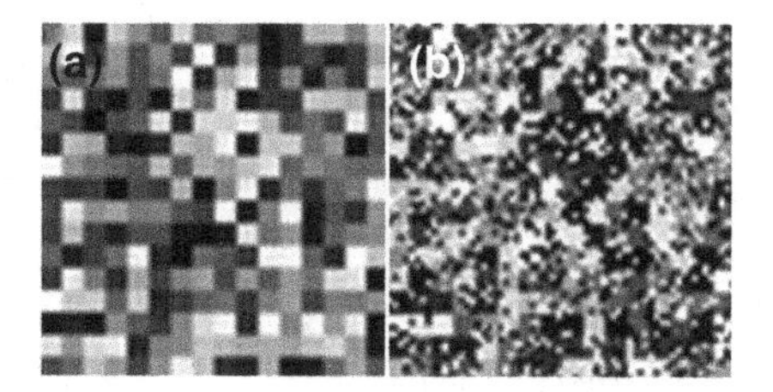

Fig. 4 Artificial training images: **a** Original image (Target training image in Fig. 3); **b** Impulse noise image (Input training image in Fig. 3)

When we have the training images shown in Fig. 4, the input–output training patterns can be obtained easily. As to each pixel of the input training image shown in Fig. 4b, the four inputs of each NF filter, x_1, x_2, x_3, and x_4, can be obtained by employing a corresponding selecting data block, respectively, according to the selecting data method described in Sect. 2.1. As to each pixel of the input training image, we also can get the corresponding target pixel in the target training image shown in Fig. 4a. This procedure can be separately repeated for all the pixels of the input training image. In this way, we can get the input–output training patterns according to the training images.

As we mentioned in Sect. 2.3, the initial value of the antecedent parameters (a_{ij}, b_{ij}, and c_{ij}) can be determined by the input–output training patterns. Each of the four NF filters has 4-inputs. Each input has three *generalized bell* type membership functions. According to the selecting data method described in Sect. 2.1, the luminance values of the inputs x_1, x_2 and x_3 range between -255 and 255, while those of the input x_4 range between 0 and 255. So the generalized bell functions shown in Fig. 5a can be employed as the input membership functions for the inputs x_1, x_2 and x_3 before training, while those shown in Fig. 5b can be used as the input membership functions for the input x_4 before training. Note that the luminance values of the inputs x_1, x_2 and x_3 are normalized between -1 and 1, while those of the input x_4 are normalized between 0 and 1. Thus we can determine the initial value of the antecedent parameters (a_{ij}, b_{ij}, and c_{ij}).

The four NF filters shown in Fig. 1 are trained individually. The learning algorithm of the each NF filter can be summarized by the following steps. Assuming that the input–output training patterns have been obtained and the initial value of the antecedent parameters (a_{ij}, b_{ij}, and c_{ij}) have been determined.

Step 1. Set $n = 1$, where n denotes the learning epoch.

Step 2. The consequent parameters d_{kl} are updated using the least-squares optimization algorithm described in Sect. 2.3.

Step 3. The outputs of the NF filter are calculated using the new consequent parameters and the result of the objective error function can be obtained.

Step 4. If the result of the objective error function is less than the predefined threshold or the maximum number of learning epoch is reached, the learning procedure stops. Otherwise, the error signals are propagated back and the antecedent parameters (a_{ij}, b_{ij}, and c_{ij}) are updated using the gradient descent algorithm described in Sect. 2.3.

Step 5. Set $n = n+1$ and go to Step 2.

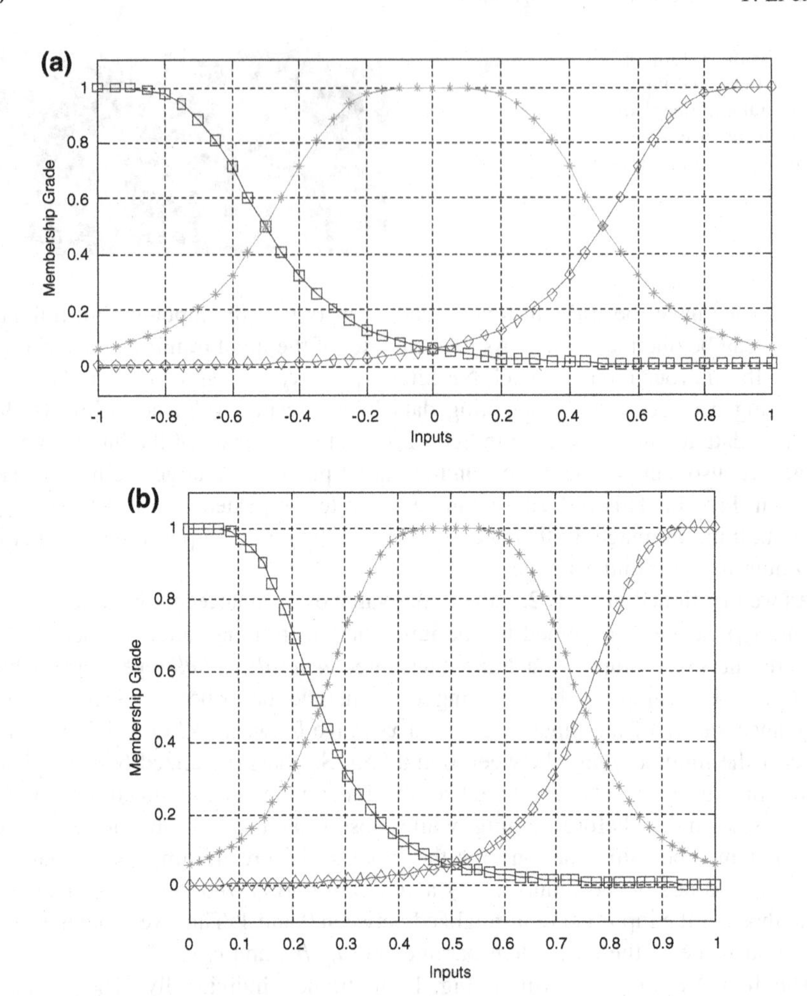

Fig. 5 Input membership functions before training. **a** Three input membership functions for the inputs x1, x2, and x3 before training. **b** Three input membership functions for the input x4 before training

2.5 Testing (Filtering) Procedure

Once the training of the four NF filters are all completed, the antecedent parameters (a_{ij}, b_{ij}, and c_{ij}) and the consequent parameters d_{kl} are fixed. Then the four NF filters can be combined with the postprocessor to construct the NF network shown in Fig. 1. The network can be applied as a noise filter for restoring gray scale images corrupted by impulse noise. In the testing (filtering) procedure, the restoration of the noisy input image can be described as follows:

Step 1. As to each pixel of the noisy input image, the four inputs of each NF filter, x_1, x_2, x_3, and x_4, can be obtained by employing a corresponding selecting data block, respectively, according to the selecting data method described in Sect. 2.1.

Step 2. Each NF filter individually generates an output value for the current operating pixel and the four outputs of the NF filters are fed to the postprocessor, which calculates the final output of the NF network by using the method described in Sect. 2.2. The final output represents the restored value of the pixel in the output image corresponding to the current operating pixel.

Step 3. The filtering procedure is repeated until all the pixels of the noisy input image are restored. Then the output image shown in Fig. 1 is the restored image.

3 Experimental Results

Extensive experimental results are presented in this section to demonstrate the performance of the proposed impulse noise filter. In our simulation, the eight original images are 8-bit gray scale images having the same size of 256 × 256 pixels. These images are shown in Fig. 6. The test images are generated by corrupting the original images by impulse noise ranging from 3 to 80 %.

For comparison, the corrupted test images are also filtered by using several conventional filters, such as, the *standard median filter (SMF)* [1], *edge-detecting median filter* (EDMF) [4], *multi-state median filter* (MSMF) [5], *signal-dependent rand-ordered mean filter* (SDROMF) [6] and *fuzzy filter* (FF) [7]. Note that the conventional filters EDMF, MSMF, SDROMF and FF have a number of tuning parameters. The values of these parameters are heuristically determined. In our experiments, the values suggested in the corresponding references are used.

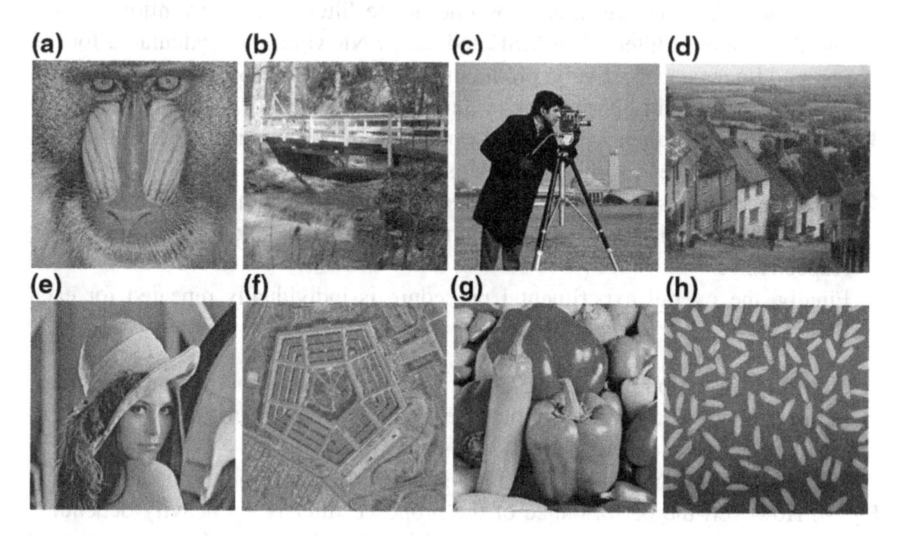

Fig. 6 Testing images: **a** Baboon; **b** Bridge; **c** Cameraman; **d** Goldhill; **e** Lena; **f** Pentagon; **g** Peppers; **h** Rice

For EDMF, $T = 116$. For MSMF, $w_{\max} = 5$ and $T = 20$. For SDROMF, $\alpha_1 = 0$, $\alpha_2 = 1$, and the thresholds $\{T_1, T_2, T_3, T_4\}$ are $\{8, 20, 40, 50\}$. For FF, $L = 256$, $a = 40$ and $b = 32$.

Both quantitative and qualitative evaluations have been presented to demonstrate the superior filtering performances of the proposed filter.

3.1 Quantitative Evaluation

Two quantitative measures, *mean squared error* (MSE) and *peak signal-to-noise ratio* (PSNR), are used to assess the filtering performance. They are defined as

$$MSE = \frac{1}{M \times N} \sum_{i=1}^{M} \sum_{j=1}^{N} (O(i,j) - R(i,j))^2 \tag{17}$$

$$PSNR = 20 \log_{10} \left(\frac{255}{\sqrt{MSE}} \right) \tag{18}$$

where $O(i,j)$ and $R(i,j)$ represent pixels of the original and the restored versions of a corrupted test image with the size of $M \times N$, respectively. Note that the smaller value for MSE and the bigger value for PSNR mean better filtering performance.

The experimental procedure can be described as follows.

1. For each noise density, the eight images shown in Fig. 6 are corrupted by impulse noise with that noise density. So we can get eight different noisy images each having the same noise density.
2. These noise images are filtered by one of the filters (five conventional filters and the proposed filter). The MSE and the PSNR values are calculated for the restored images, respectively, producing eight different MSE and PSNR values, which represent the filtering performance of that filter for that noise density.
3. These values are then averaged to obtain the representative MSE and PSNR values of that filter for that noise density, respectively.
4. This procedure is separately repeated for all noise densities to obtain the average MSE and PSNR values of that filter as a function of noise density, respectively.
5. Finally, the overall experimental procedure is individually repeated for each filter.

The results are reported in Fig. 7. It is easy to conclude that the proposed filter, in terms of the two quantitative measures, clearly outperforms the five conventional filters for all noise densities.

Note that, in our simulation, we use a simple artificial training image shown in Fig. 4. However, the performance of the proposed filter is not heavily dependent on the training image. That is, the proposed filter is still able to obtain a very good filtering performance, even though the testing images are not as same as the

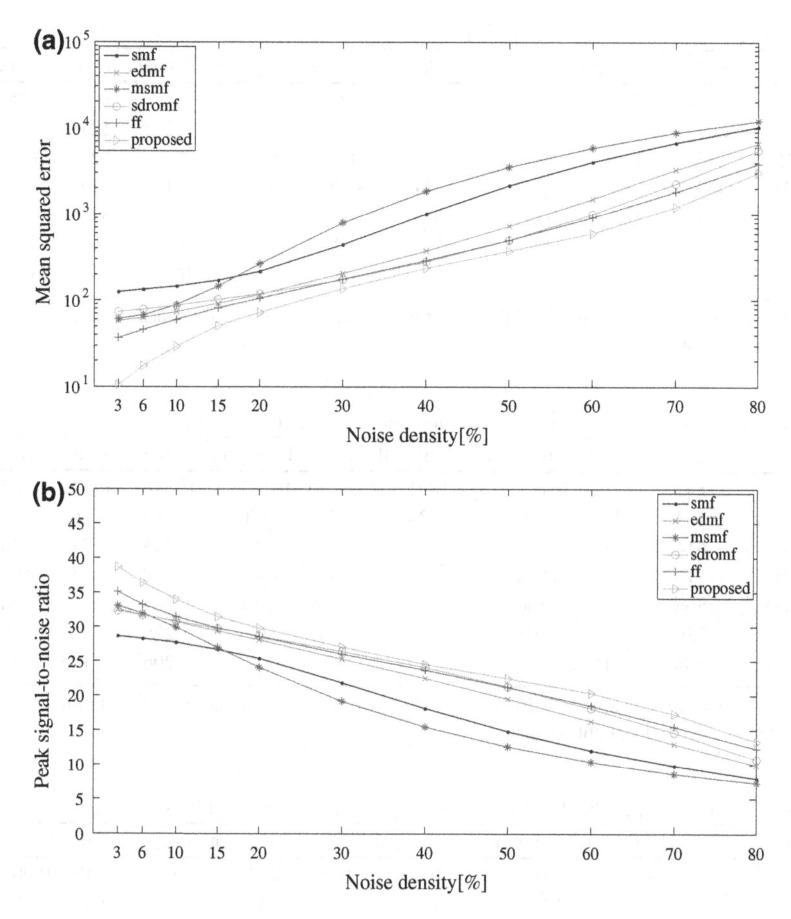

Fig. 7 Filtering performance of the proposed filter compared with the conventional filters. The testing images are corrupted by impulse noise ranging from 3 to 80 %. **a** Mean squared error (MSE); **b** Peak signal-to-noise ratio (PSNR)

training image and the noise density of the testing images is different from that of the input training image. Therefore, the proposed filter exhibits a satisfactory performance in robustness. During training, we just use a simple artificial training image regardless of what the testing image is and what the noise density is.

In order to separately analyze the noise suppression and detail preservation performance of all filters, the MSE values are calculated (1) for all pixels, (2) for corrupted pixels only and (3) for uncorrupted pixels only. These values are given in Tables 2, 4, respectively. The test images are all corrupted by 20 % impulse noise.

The comparison of the overall filtering performances of the filters is described in Table 2. The MSE values are calculated for all pixels of the input images. It is clear that the proposed filter exhibits the best performance for all test images.

The MSE values listed in Table 3 are calculated for only the corrupted pixels of the input images. These values show the comparison of the noise suppression

Table 2 Comparison of the overall filtering performances of the filters

Filter	Baboon	Bridge	Cameraman	Goldhill	Lena	Pentagon	Peppers	Rice	Average
Noisy	3519	3866	4041	3789	3692	3825	3816	3677	3778
SMF	501	261	238	182	106	205	112	121	216
EDMF	225	148	188	89	60	104	69	52	117
MSMF	447	288	294	241	190	256	201	194	264
SDROMF	321	143	197	81	37	98	48	32	119
FF	217	118	180	79	62	84	58	48	106
Proposed	132	97	86	61	43	79	46	42	73

The test images are corrupted by 20 % impulse noise. The MSE values are calculated for all pixels of the input images

Table 3 Comparison of the noise suppression performances of the filters

Filter	Baboon	Bridge	Cameraman	Goldhill	Lena	Pentagon	Peppers	Rice	Average
Noisy	17542	19317	20169	18918	18490	19157	19088	18333	18877
SMF	734	452	433	342	230	376	249	264	385
EDMF	614	488	455	322	228	364	245	217	367
MSMF	1373	1187	1198	1042	898	1083	958	936	1084
SDROMF	566	356	366	206	118	251	144	115	265
FF	589	449	401	320	261	346	225	221	352
Proposed	604	415	357	273	194	340	206	195	323

The test images are corrupted by 20 % impulse noise. The MSE values are calculated for only the corrupted pixels of the input images

Table 4 Comparison of the detail-preservation performances of the filters

Filter	Baboon	Bridge	Cameraman	Goldhill	Lena	Pentagon	Peppers	Rice	Average
Noisy	0.00	0.00	0.00	0.00	0.00	0.00	0.00	0.00	0.00
SMF	442	213	190	142	74	163	78	86	173
EDMF	127	63	121	322	18	39	25	10	91
MSMF	215	64	68	41	13	50	12	8	59
SDROMF	260	89	154	49	17	60	24	11	83
FF	124	36	125	19	12	18	16	5	44
Proposed	13	17	18	8	5	14	7	4	11

The test images are corrupted by 20 % impulse noise. The MSE values are calculated for only the uncorrupted pixels of the input images

performances of the filters. It is obvious that the noise removal performances of the proposed filter are better than the other filters except the SDROMF.

Table 4 lists the MSE values calculated for only the uncorrupted pixels of the input images. It denotes the comparison of the detail-preservation performances of the filters. In the case of impulse noise removal, the aim of optimal filtering is to design noise reduction algorithms that would affect only corrupted image pixels, whereas the undistorted image pixels should be invariant under the filtering operation. Therefore, the less of the MSE value calculated for only the uncorrupted pixels of a test image means the better detail-preservation performance of the filter.

It can be easily seen from Table 4 that the MSE values of the proposed filter are significantly lower than those of the other filters and much closer to zero. Hence, the detail-preservation performances of the proposed filter are considerably better than other conventional filters.

3.2 Qualitative Evaluation

In addition to the quantitative evaluation presented above, a qualitative evaluation is necessary since the visual assessment of the processed images is ultimately the best subjective measure of the effectiveness of any method. Figure 8 depicts the superior filtering results of the proposed filter for the Baboon image corrupted by 30 % impulsive noise in comparison with the other conventional filters. Figure 8h shows the restored image using the proposed filter. It is obvious that the proposed filter has the excellent ability of noise attenuation because there is almost no impulse noise left in the restored image shown in Fig. 8h. Compared with the restored images using other conventional filters, the proposed filter also has the desirable capability of details preservation as indicated especially on the beard part of the Baboon image.

From the experimental results presented above, it can be easily seen that the proposed new filter provides significantly good results in the testing images corrupted by different percentages of impulse noise, and outperforms the other conventional filters under consideration.

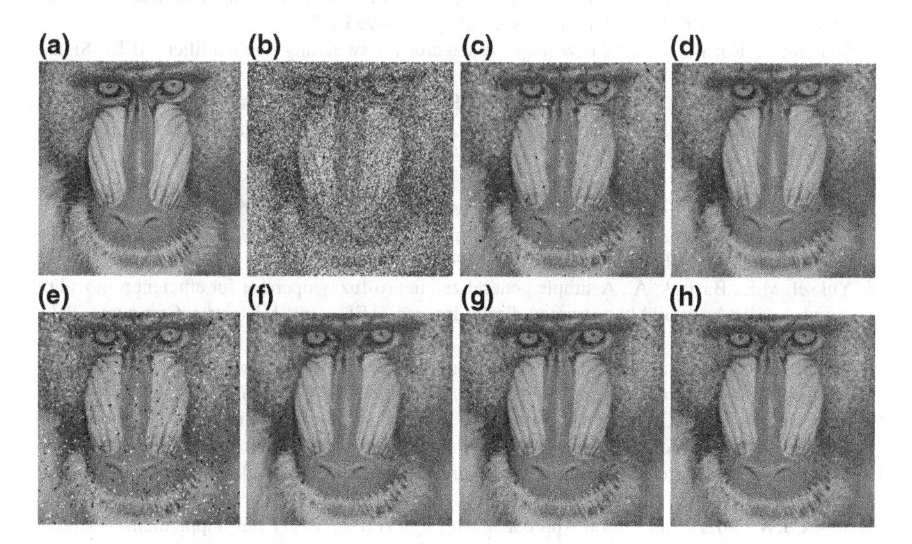

Fig. 8 Filtering results for the testing image Baboon. **a** Original image. **b** Image corrupted by 30 % impulse noise. **c** Restored image using SMF. **d** Restored image using EDMF. **e** Restored image using MSMF. **f** Restored image using SDROMF. **g** Restored image using FF. **h** Restored image using proposed filter

4 Conclusion

A novel impulse noise filter for gray scale images based on a neuro-fuzzy network is presented. The network is constructed by combining four NF filters with a postprocessor. As supported by the simulation results, the proposed filter compares favorably with conventional techniques in the capabilities of noise attenuation and details preservation, in both quantitative and qualitative measures. The advantages of the proposed filter can be summarized as follows:

1. The proposed filter has no tuning parameters.
2. During the training procedure, the hybrid learning rule is applied to reduce the error to improve the speed and the performance of the optimization.
3. A simple artificial image is used during training procedure regardless of what the testing image is and what the noise density is. Therefore, the proposed new filter exhibits a satisfactory performance in robustness.

References

1. Pratt, W.K.: Digital Image Processing. Wiley Interscience, New York (1978)
2. Yli-Harja, O., Astola, J., Neuvo, Y.: Analysis of the properties of median and weighted median filters using threshold logic and stack filter representation. IEEE Trans. Signal Process. **39**(2), 395–410 (1991)
3. Ko, S.J., Lee, Y.H.: Center weighted median filters and their applications to image enhancement. IEEE Trans. Circuits Syst. **38**(9), 984–993 (1991)
4. Shuqun, Z., Karim, M.A.: A new impulse detector for switching median filters. IEEE Signal Process. Lett. **9**(11), 360–363 (2002)
5. Tao, C., Hong Ren, W.: Space variant median filters for the restoration of impulse noise corrupted images. IEEE Trans. Circuits Syst. II Analog Digital Signal Process. 2001. **48**(8), 784–789
6. Abreu, E., et al.: A new efficient approach for the removal of impulse noise from highly corrupted images. IEEE Trans. Image Process. **5**(6), 1012–1025 (1996)
7. Russo, F., Ramponi, G.: A fuzzy filter for images corrupted by impulse noise. Signal Processing Letters, IEEE **3**(6), 168–170 (1996)
8. Yuksel, M.E., Basturk, A.: A simple generalized neuro-fuzzy operator for efficient removal of impulse noise from highly corrupted digital images. AEU—Int. J. Electron.Commun. **59**(1), 1–7 (2005)
9. Li, Y., Chung, F.-L., Wang, S.: A robust neuro-fuzzy network approach to impulse noise filtering for color images. Appl. Soft Comput. **8**(2), 872–884 (2008)
10. Jang, J.-S.R., Sun, C.-T.: Neuro-fuzzy and soft computing: a computational approach to learning and machine intelligence, p. 614. Prentice-Hall, Inc, Upper Saddle River (1997)
11. Masters, T.: Advanced Algorithms for Neural Networks. John Wiley & Sons, New York (1995)
12. Hines, J.W.: Fuzzy and Neural Approaches in Engineering, MATLAB Supplement. Adaptive and Learning Systems for Signal Processing, Communications and Control Series. In: Haykin, S. (ed.) New York, John Wiley and Sons (1997)

A Fuzzy System for Educational Tasks for Children with Reading and Writing Disabilities

Adalberto Bosco C. Pereira, Gilberto Nerino de Souza Jr., Dionne C. Monteiro and Leonardo B. Marques

Abstract This paper proposes a computational approach that aims the use of Artificial Intelligence in Education (AIED) to aid teachers, psychologists and educationalists in the learning process of reading. This approach aims at generating teaching tasks which can be accordingly adapted to the individual needs of each student. While the tasks are being executed, a Machine Learning (ML) system will collect and process data to allow an analysis of the student's learning process for each individual word in reading and writing abilities. The fuzzy system will propose an appropriate task based on the data collected by the ML. The output of the fuzzy system is an adapted task that will motivate the children in execution of the tasks.

Keywords Artificial intelligence · Fuzzy · Machine learning · AIED · Teaching tasks

A. B. C. Pereira (✉) · G. N. de Souza Jr. · D. C. Monteiro
Laboratory of Applied Artificial Intelligence, Institute of Exact and Natural Sciences—Federal University of Para, Para, Brazil
e-mail: adalberto@ufpa.com

G. N. de Souza Jr.
e-mail: gilbertojr@ufpa.com

D. C. Monteiro
e-mail: dionne@ufpa.com

L. B. Marques
Laboratory for the Study of Human Behavior, Federal University of So Carlos, So Carlos, Brazil
e-mail: leobmarques@gmail.com

I. Hatzilygeroudis and V. Palade (eds.), *Combinations of Intelligent Methods and Applications*, Smart Innovation, Systems and Technologies 23,
DOI: 10.1007/978-3-642-36651-2_4,

1 Introduction

Digital games have an important space in the lives of children, teenagers, and adults. It is one of the fastest growing sectors in the media and entertainment industry. In addition, games have been combined with a wide range of fields such as unusual simulations, Games for Health, Game-Based Learning (GBL) [1] and Artificial Intelligence in Education (AIED) [2].

Good learning requires teachers and students to combine their efforts. In the case of the students, their interest in studying is of fundamental importance [3] to ensure they can achieve their maximum potential. For the teachers, it can be concluded that any attempt to improve students' achievements must be based on the acquisition of an effective teaching behavior [4]. In other words, they should give students appropriate guidance.

Usually gamification is applied to non-game applications and processes to motivate people to adopt them or to influence how they are used. Gamification works for: (i) making technology more engaging, (ii) encouraging users to engage in desired behaviors, (iii) showing a path to mastery and autonomy, (iv) helping to solve problems and not being a distraction, and (v) taking advantage of humans' psychological predisposition to engage in gaming. The technique can encourage people to perform tasks that they ordinarily consider tedious, such as completing surveys, shopping, filling out tax forms, or reading web sites.

For several years, some games have incorporated automatic generation to create levels, missions, and spaces that are designed to increase the lifetime of their games. There are games that have a form of challenge called "endless", where the player is exposed to challenges at levels that are generated indefinitely until the game is lost. This approach is adopted for several areas in games and in many features such as: (i) 2D textures, (ii) 3D models, (iii) music, (iv) levels, (v) story, (vi) mission, and others. These studies are of increasing importance in the process of developing computer games [5].

The purpose of this work is to test a Fuzzy Logic solution applied to the current task of teaching reading in a digital game that is adaptive to the individual needs of each player. That is, if the new tasks of education are suited to pre-existing literacy skills of the student, this means that this task must not be either too easy or too difficult. The data collected during the execution of the teaching tasks will be preprocessed and analyzed with the aid of a Machine Learning (ML) system, which will provide data to Fuzzy System to evaluate and consider the best choice for proposing a task. The output of the Fuzzy System is determined by the data required for task generation, which in turn will be transformed into an adapted level of the game.

2 Related Work

The "*Gerenciador de Ensino Individualizado por Computador*" (GEIC—Computer Individualized Education Manager) [6] approaches the problem of ensuring the dynamic generation of content for educational use, providing by software the programming procedures based on choice teaching tasks. It allows the creation of Teaching Units that combine various teaching tasks, and represent discrete attempts to provide choice tasks. The objective of this work is to aid teachers in teaching reading and writing abilities to children with learning difficulties. Further details of this program will be given in section Three of this paper.

Azevedo and Marques [7] discusses methods of teaching which, although of satisfactory standard, still present different efficiency levels for each student. This is discussed in examining the computational and educational resources and their degree of adaptability to the individual needs of each student.

A study in the field of AIED, Benedict [2], state that the individualization of teaching instructions can be effective and is regarded by the author as the "Holy Grail of AIED". This paper compares the educational differences in AI systems (AIED) with conventional educational systems used in the classroom or traditional methods of Computer-Assisted Instruction (CAI).

The learning program called GEIC [6], was transformed into a game called ALE-RPG [8]. However, it should be stressed that the structured progress that the game kept in the GEIC proved to be a little static. The progression of the player through the teaching tasks does not allow a fully customized advance to be made, since it is conditioned by the need to learn all the words of each teaching unit. The non-acquisition of any word component of the teaching units delays the teaching progress of the next units.

There are games like Diablo [9], Torchlight [10], Spore [11] and MineCraft [12] that use automatic generation levels when the player starts a new game, but keeps the missions and game objectives distributed randomly on the generated map. However, all this randomness follows 'brute force' algorithms, without any predetermined scale of difficulty.

The approach of Dormans [5] investigates strategies to generate levels of action-adventure games that are divided into two individual structures, so that they generate missions first and then spaces. The different types of generative grammar are analyzed in a search for the one that best fits.

Xiangfeng [1] uses Fuzzy Cognitive Maps to design Game-Based Learning (GBL). The goal is to use the Hebbian learning rule to increase learning capacity by employing the game data and Unbalance Degree to establish the lack of prior knowledge.

3 Teaching Program

Since the 1980s, researchers in Brazil in the field of behavioral psychology have been refining a program to help children with a previous history of school failure to learn to read. This procedure, called "*Aprendendo a Ler e a Escrever em Pequenos Passos*" (ALEPP—Learning to Read and Write in Small Steps) [13], is mainly concerned with detecting and overcoming problems that are found in children who have literacy difficulties in an efficient manner. It also provides tasks that teach the basic components of reading in a personalized manner.

The GEIC is a remote software that allows the ALEPP curriculum [13] to be applied. However, it is not yet adapted enough because it is composed of static tasks, grouped into Teaching Units that have been previously determined by specialists.

The Matching to Sample (MTS) procedure is used for teaching reading relations programmed into ALEPP [14] through GEIC [6], Fig. 1. This procedure is used to teach relations between printed words, pictures and dictated words. In this task, one stimulus (the sample) must be matched to the correct comparison.

Each task contains *n* comparisons, ranging from one to three, where one acts as a stimulus by referring to the sample on top of the screen. Finally, the last feature is the definition of the words that correspond to the model stimulus and stimuli choices. The tasks are subdivided into types of stimuli combinations such as AB, BC and CB. In Fig. 1 the sample stimulus is the sound that corresponds to the correct alternative.

The tasks allow you to create relationships between the stimuli of different modalities. They are relations that are taught between the dictated words (A), image representation of this word (B) and the printed word (C). The tasks that establish these relations are listed as tasks of type AB (dictated word-picture), CB (printed word-picture) and BC (picture-printed word). All relationships of all tasks types for reading are described in Table 1.

There are also tasks that are characteristic of word construction, where the correct model is presented and the child needs to build the word syllable by syllable, or letter by letter, that is equivalent to the correct model, whether it is a figure type stimuli, sound stimuli or text type stimuli. For example, in attempts like

Fig. 1 Task type AB

Table 1 Illustrations of teaching tasks related to reading acquisition

Task type	Role of the child
AB	Given the spoken statement "Point at Student" the student must select the student figure
AC	Given the spoken statement "Point at Cat" the student must select the word 'cat'
BC	Given the figure of a ham mock, the student must select the word "ham mock"
CB	Faced with the word "candle", the student must select the candle figure

Table 2 Illustrations of teaching tasks related to the acquisition of writing

Task type	Role of the child
AE	Given the spoken statement "Type Hammock", the student should write the word "hammock" by choosing the letters in the correct order
BE	Given the spoken statement "What figure is this?", the student should write the word "lock" by choosing the letters in the correct order
CE	Given the spoken statement "What word is this?", the student should write the word "lock" by choosing the letters in the correct order

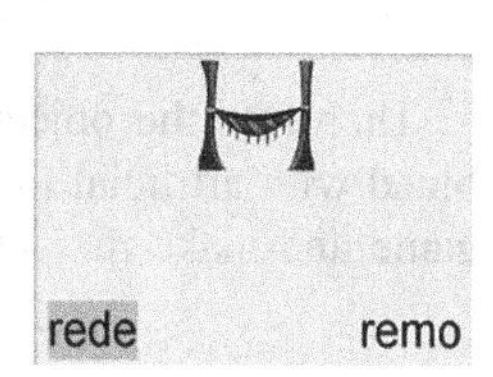

Fig. 2 Task type BV

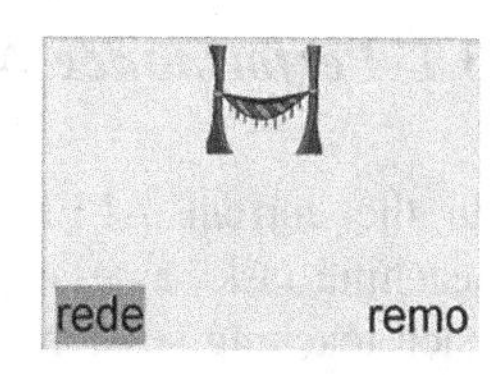

Fig. 3 Task type BC

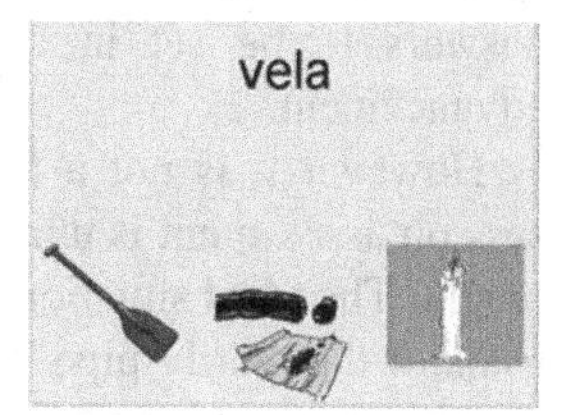

Fig. 4 Task type CB

the BE type. The model stimulus is a picture and the choices are scrambled syllables where there is at least one sequence of syllables of the correct word. The student must select the syllables that correspond to the model in the correct order. Table 2 displays tasks for writing acquisition (Figs. 2, 3, 4, 5, 6 and 7).

Fig. 5 Task type AE

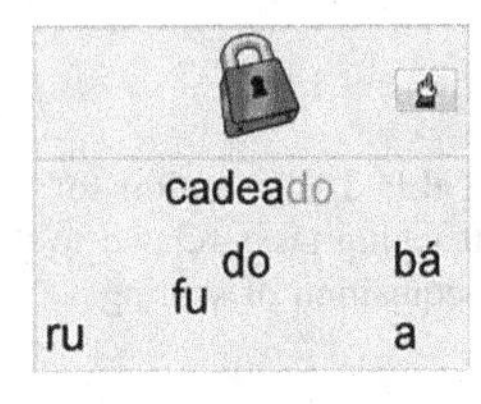

Fig. 6 Task type BE

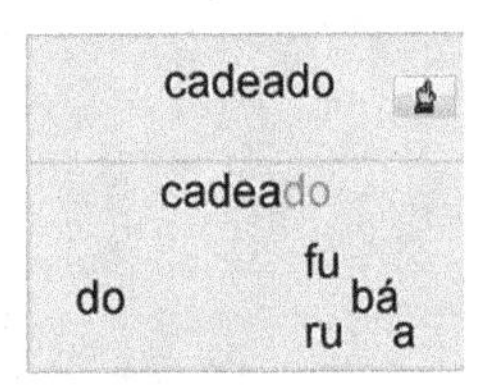

Fig. 7 Task type CE

Therefore the objective of this work is to use this educational program combined with artificial intelligence to classify and generate computational tasks in a game and make the activity more entertaining and interesting for students.

3.1 Performance Assessment in Reading

In the current ALEPP assessment system, the GEIC automatically loads the teaching tasks assessments [15]. The GEIC can be set individually at the end of each teaching session, whether the student has repeated the last block of tasks or advanced to the next block. This software allows a minimum number of correct answers to be set in a session and uses these successes as a criterion for advancement.

However it is not a fine evaluation in terms of the words learned. The final session assessment is unable to identify the different kinds of errors for each word taught. The ideal situation is to evaluate correctness for each word. For example, in trials of the word "pipe" as a sample, it shows that the student learned the word, how he knows this word, and then a task is set with a focus on learning how to read or write, so that the student can learn to read or write that word.

4 Our Approach

In order to adopt an approach to assist instructors, the general purpose of the project is initially aimed at creating one AIED to teach reading and writing and then incorporate it into a digital game. AIED is composed of two intelligent systems, ML and Fuzzy System, which will act together, as shown in the diagram in Fig. 8. They represent the Machine Learning system that is not covered in this work, but developed separately in parallel [16]. Both systems will be executed during game-play of several mini-games.

The interaction between the student and the game will be initially by computers, using mouse and keyboard. In the future, we intended to use tablets with the Android operational system. Both systems (computers and tablets) will be composed of several mini-games with gamefied teaching duties.

4.1 Game Story

The game story consists of the problems that an alien acquires when he accidentally falls on our planet. Coming from a distant planet in his spaceship, the main character is a traveler who seeks to explore new planets and learn about them.

To continue his journey, he must have some pieces in order to repair the ship. Our character, named Amaru (Fig. 9), needs to learn to communicate by reading and writing in Earths language to proceed with his interplanetary exploration. During his journey he learns to communicate and gets to know humans that will

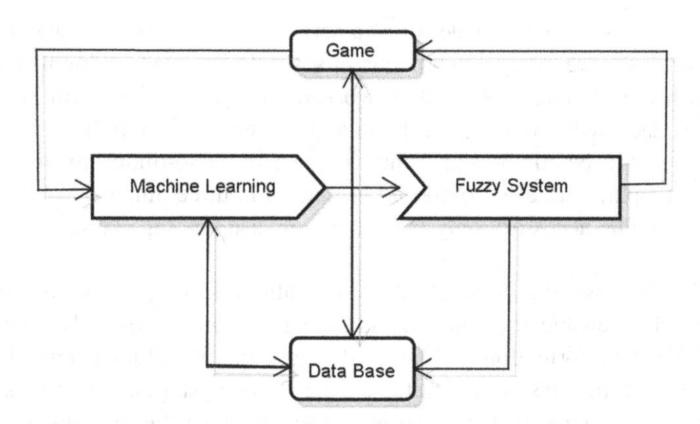

Fig. 8 Macro view of the project

Fig. 9 Amaru

Fig. 10 Urama

help him get the parts of his ship. He will also have help from his robot Urama (Fig. 10), an NPC (Non Player Character), which will help to solve the teaching tasks of the game.

4.2 Game Mechanic

The platform-style game is made up of a series of different challenges, in which the teaching tasks are docked and occur in several different environments on the planet. Each type of challenge will be worked on separately and will be called a mini-game. Each will have its own logic and mechanical solution.

The player will be able to move the character in a two-dimensional plane, but the game's graphics are also composed of objects in three dimensions. To control the game it will be necessary only the use of a mouse, the Fig. 11 shows the game runing.

The game begins with an initial pretest of static tasks to generate the minimum data required to enable the Machine Learning system to correctly analyze the student. After the information is obtained during game-play, for each word that has been processed a new task is generated by the Fuzzy System and will be stored in the data-base. The level of the game is created by using the features of the generated task. The sequence of tasks follows that of Logical Teaching. For the specialists, there is a preferred order for each word being taught, and new words will enter the teaching tasks gradually, depending on the degree of literacy of the word.

Fig. 11 Gameplay

The ML is responsible for evaluating the student, calculating his knowledge level, and providing the probability of success and difficulty of a given task. Done so, such data is passed to the Fuzzy System in order that it can correctly generate an adapted teaching task.

For each Teaching Unit there are fifteen words that have to be taught. The ML will decide the level of knowledge and evaluate if the student has learned every single word. The ML will run for each new generated task, and it will always be calculated on the basis of this new information regarding the degree of knowledge of a particular word. The words that are determined as literacy appear less frequently, and when every word has been learned, the game is over. Figure 12 shows a diagram of the operation of this proposal.

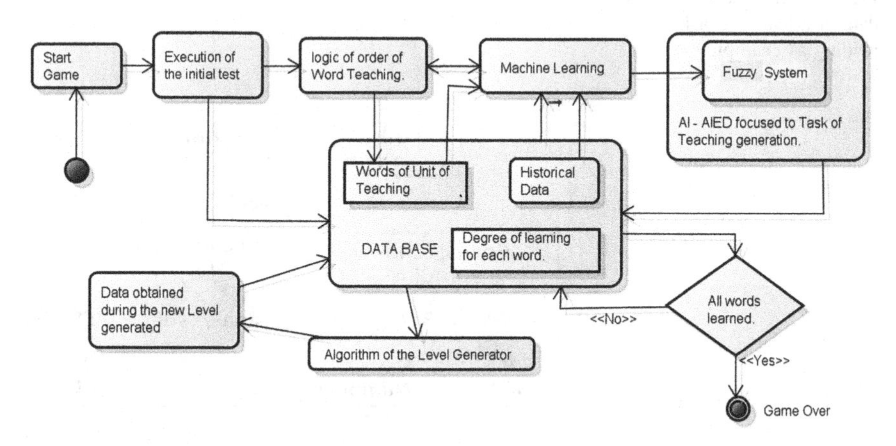

Fig. 12 Flow diagram of the approach

5 Artificial Intelligence Systems

This session aims to describe the artificial intelligence used in the game. The part about the ML is discussed briefly because it is treated and described in parallel work [16] of the same project.

5.1 Machine Learning

In this work the Machine Learning's goal is to estimate a degree of learning of reading and writing of individual words, using it to analyze the behavior of the student during the teaching session. This aim proposes the word's learning and the students understanding of the task. In essence, the supervised learning is used to perform the prediction of student learning and to classify a level of knowledge of a specific word. It is applied for each word model in order to define a standard of the students learning [16]. Basically, for the AI system, a teaching task is a set of data as shown in Fig. 13.

One of the extracted information of schematic shown above is equivalent to the difficulty of the task. To determine the learning characteristics of the student in the system, two equations were defined in order to figure out how much a learning task is difficult for the didactic program in focus, scoring the factors of task difficulty [16]. All equations were based on attributes discussed in conjunction with psychologists, i.e., in each element of the equations was considered one feature of difficulties which can increase or decrease the difficulty of tasks, according to psychologists. The Eqs. 1 and 2 that describe the difficulty of the reading task and the difficulty of the writing task, respectively, are shown below.

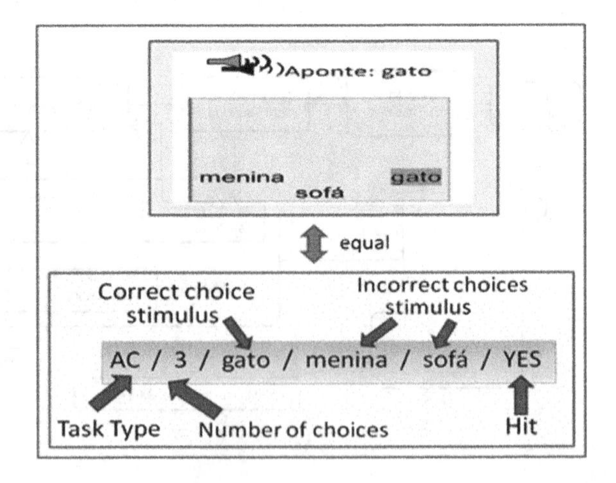

Fig. 13 Schematic of various data of a learning task to reading as example

$$Dr = \begin{cases} \dfrac{\sum_{i=1}^{n-1} P_{in}}{n-1} * \alpha + T + \beta + \dfrac{n}{n_{max}} * \gamma, & n >; 1 \\ T * \beta + \dfrac{n}{n_{max}} * \gamma, & n = 1 \end{cases} \quad (1)$$

Where,

D_r	: Difficulty of the reading task,
P_{in}	: proximity factor to incorrect choices in relation to the model word,
T	: task type score,
n	: number of choices available in the task,
n_{max}	: maximum of possible choice stimuli,
α, β and γ	: difficulty factors score, obtained empirically.

$$Dw = \frac{S_i}{S_t} * \alpha + T * \beta \frac{S_t}{T_{sm}} * \sigma \quad (2)$$

Where,

D_w	: Difficulty of the writing task,
S_i	: number of incorrect choice of syllables over the word model,
S_t	: total number of choices in the task,
T	: type of writing task score,
S_{max}	: total number of syllables possible for a task,
S_m	: total number of choices or syllables of the word model of the task,
T_{sm}	: total number of choices or syllables of the word with the greatest amount of syllables system,
α, β, γ and σ	: score for the factors of difficulty regarding the writing tasks obtained empirically.

5.2 Description of the Fuzzy System

The main objective of this paper is to make use of data generated by ML correctly in order to generate an adjusted task. Fuzzy Logic was chosen to carry this out for the following reasons: it has a distinct capacity to express the vagueness and uncertainty of the knowledge it represents [17]; it is able to model a system close to logical grammatical rules; it ensures a better approximation to the knowledge of

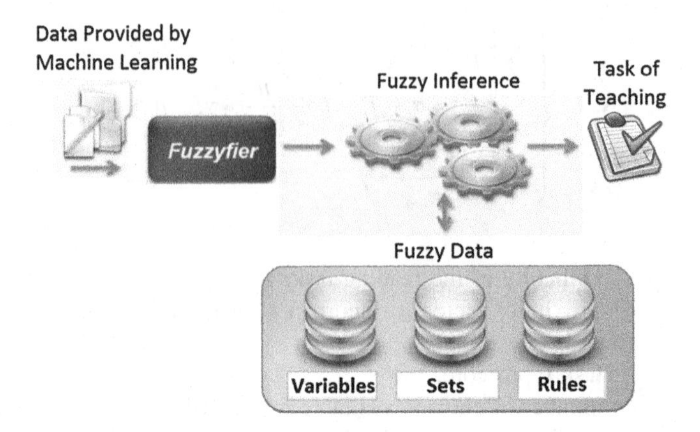

Fig. 14 Macro view of the Fuzzy System

a specialist through semantic representations and linguistic terms; and it operates by choosing few rules and working with imprecise terms [18].

The data provided by ML are fuzzified and separated into fuzzy groups. For each task feature, one fuzzy inference occurs. At the end of the system, the output is the completion of the task, as shown in Fig. 14.

Fuzzification. All the data processed and generated by ML, are abstracted and normalized in three fuzzy sets corresponding to a numerical range, from 0 to 100 % and a degree of pertinence ranging from 0 to 100 % [18], as shown in Fig. 15. The fuzzy values represented in this approach correspond to the descriptors for the trapezoidal and triangular functions illustrated in Fig. 15. In the present work all the fuzzy sets have these same values of classification. The representation of fuzzy set partitioning for a linguistic variable, showing the vertical and horizontal axes referring to the degree of pertinence and fuzzy values, respectively.

This Fuzzification is done individually for each input variable of a Task Type, Number of Comparisons and Incorrect Words of a particular model of word, described in Table 3.

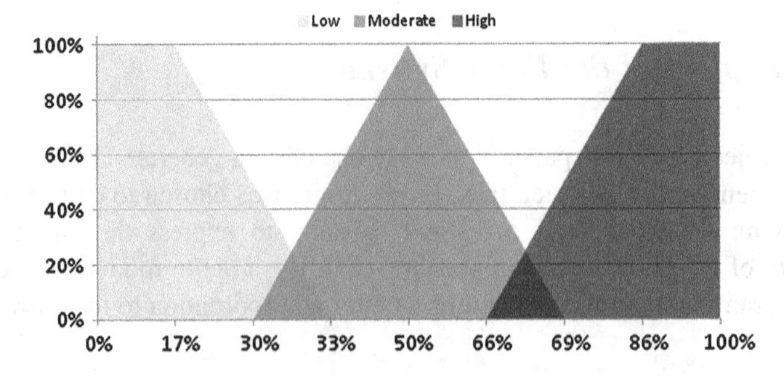

Fig. 15 Graphical representation of fuzzy set

Table 3 Input variables of the fuzzy system

Input variable
PTT: probability of hit with determined task type
TTT: hit rate of task type
PNC: probability of hit with determined number of comparisons
TNC: hit rate of number of comparisons
PPI: probability of hit with determined incorrect word
TPI: hit rate with determined incorrect word

PTT is the variable referring to the probability of success in a particular task type, which is analyzed in conjunction with TTT (hit rate of one task type). It determines whether the teaching task is generated for writing or reading, if it is of an easier or difficult type, and still if it is not of the same type repeated times.

PNC is the variable responsible for measuring the probability of success of a certain number of incorrect comparisons in order to adequately modify the degree of difficulty for the player, leaving the game more dynamic and non repetitive. This variable is analyzed in conjunction with the TNC (hit rate of n tasks with incorrect comparisons).

The probability of success with of determined incorrect word is called PPI. This variable is responsible for checking the probability of success when a specific word appears in the task as an incorrect stimulus and it is analyzed in conjunction with the hit rate of a given incorrect word (TPI). Thus, it is possible to adjust the task with the ideal incorrect comparisons in order to generate an adapted difficulty and prevent that the same task is generated repeated times.

The above inputs used in the Fuzzy System are the outputs of the Machine Learning.

Rule Sets. The architecture of Fuzzy Logic provided here is mapped out in a set of rules reflecting the ideas of specialists about the Study of Human Behavior project [19], as well as interviews that were conducted for this study and the works outlined above.

The set of rules for this work is designed to create the appropriate learning tasks in an efficient manner, and also aims at encouraging the students, ensuring the game is kept enjoyable and stimulating.

If the tasks that are generated are too difficult for the player and he starts to miss too much, the player will not learn and may be discouraged and lose interest in the game. On the other hand, if there is too little difficulty for a player, he may also become discouraged, and will not be able to play the game to its full potential, and thus be mining the teaching process, which can delay his learning. From this perspective, the aim is to reach a balance between degrees of difficulty for each student.

The rules that are set (with the assistance of specialists) attempt to treat the input variables so that the choices of output variables, which are the characteristics of the new task, have an appropriate degree of difficulty. In addition, a further aspect of this logic is to ensure that the same feature does not appear too often and, therefore, prevents the tasks from becoming repetitive.

Table 4 Output variables

Output variable
DTT: need for task type
DNC: need for number of comparisons
DPI: need for incorrect word

Each task performed by the ML player will pass on updated data about the student's progress, which means that the rules of the fuzzy system will always generate an appropriate task based on the updated data.

Fuzzy Inference. The fuzzy inference system used in this work uses the Mamdani model [20]. It corresponds to the algorithm of fuzzified information processed in accordance with linguistic rules, [21] which are defined by specialists and the research studies referred to. Table 4 correlates the input variables with the output in logical terms and the "if-then" form in causal terms. In the initial testing, the following rules were obtained and are arranged in Tables 5, 6, 7, 8, 9, 10, 11, 12,13 and 14 as follows: "**If** a variable in column 1 = X **and** variable in column 2 = Y **then** variable in column 3 = Z".

The inference occurs for each task type resulting in a linguistic activation and their respective pertinence. To better decide which is the best task type it is necessary to individually check each activation of each task type. The rule set defined along with the experts and their respective activated linguistic values are shown below. Afterwards, the analysis of what task type is ideal for appropriate generation of what was mentioned in the previous topic will be made. The tasks types mentioned above, analyzed by the system shall be as follows: AE, AC, AB, BC, BE, CE. The rules for the activations are described in the Tables 5, 6, 7, 8, 9, 10, 11, 12,13 and 14 below.

The activations referred to the number of comparisons are equivalent for each number, ranging from 1 to 4 comparisons at most. Table 15, below, shows these activations.

In order to correctly define which incorrect words should be part of the new generated task, each word and their appropriate activations should be individually verified in order that they have an adapted difficulty, the tasks do not become

Table 5 Activation to the task type AE

PTT: AE	TTT: AE	DTT
Low	Low	Low
Low	Moderate	Low
Low	High	Moderate
Moderate	Low	Moderate
Moderate	Moderate	High
Moderate	High	Moderate
High	Low	Moderate
High	Moderate	High
High	High	Low

Table 6 Activation to the task type AC

PTT: AE	TTT: AE	DTT
Low	Low	Low
Low	Moderate	Low
Low	High	Moderate
Moderate	Low	Moderate
Moderate	Moderate	High
Moderate	High	Moderate
High	Low	Moderate
High	Moderate	High
High	High	Low

Table 7 Activation to the task type AB

PTT: AB	TTT: AC	DTT
Low	Low	Low
Low	Moderate	Low
Low	High	Moderate
Moderate	Low	Moderate
Moderate	Moderate	High
Moderate	High	Moderate
High	Low	Moderate
High	Moderate	High
High	High	Low

Table 8 Activation to the task type BE

PTT: BE	TTT: BE	DTT
Low	Low	Low
Low	Moderate	Low
Low	High	Moderate
Moderate	Low	Moderate
Moderate	Moderate	High
Moderate	High	Moderate
High	Low	Moderate
High	Moderate	High
High	High	Low

Table 9 Activation to the task type CE

PTT: CE	TTT: CE	DTT
Low	Low	Low
Low	Moderate	Low
Low	High	Moderate
Moderate	Low	Moderate
Moderate	Moderate	High
Moderate	High	Moderate
High	Low	Moderate
High	Moderate	High
High	High	Low

Table 10 Activation to the task type AE

PTT: AC	TTT: AE	DTT
Low	Low	Low
Low	Moderate	Low
Low	High	Low
Moderate	Low	Low
Moderate	Moderate	Low
Moderate	High	Moderate
High	Low	Moderate
High	Moderate	High
High	High	Moderate

Table 11 Activation to the task type AE

PTT: BC	TTT: AE	DTT
Low	Low	Low
Low	Moderate	Low
Low	High	Low
Moderate	Low	Low
Moderate	Moderate	Low
Moderate	High	Moderate
High	Low	Moderate
High	Moderate	High
High	High	Moderate

Table 12 Activation to the task type AC

PTT: CE	TTT: AC	DTT
Low	Low	Low
Low	Moderate	Low
Low	High	Low
Moderate	Low	Low
Moderate	Moderate	Low
Moderate	High	Moderate
High	Low	Moderate
High	Moderate	High
High	High	Moderate

Table 13 Activation to the task type AC

PTT: CE	TTT: AE	DTT
Low	Low	Low
Low	Moderate	Low
Low	High	Low
Moderate	Low	Low
Moderate	Moderate	Low
Moderate	High	Moderate
High	Low	Moderate
High	Moderate	High
High	High	Moderate

Table 14 Activation to the task type AC

PTT: AC	TTT: AE	DTT
Low	Low	Low
Low	Moderate	Low
Low	High	Low
Moderate	Low	Low
Moderate	Moderate	Low
Moderate	High	Moderate
High	Low	Moderate
High	Moderate	High
High	High	Moderate

Table 15 Activations for numbers of comparisons

PNC	TNC	DNC
Low	Low	Low
Low	Moderate	Low
Low	High	Moderate
Moderate	Low	Moderate
Moderate	Moderate	High
Moderate	High	Moderate
High	Low	Moderate
High	Moderate	High
High	High	Low

Table 16 Activations referring to incorrect words

PPI	TPI	DPI
Low	Low	Low
Low	Moderate	Low
Low	High	Moderate
Moderate	Low	Moderate
Moderate	Moderate	High
Moderate	High	Moderate
High	Low	Moderate
High	Moderate	High
High	High	Low

repetitive and so that the student does not begin answer correctly by exclusion. Thus, he learns to compare the correct word with other words in different levels of difficulty. The Table 16 is the inference model for each word.

As an example that points to a rule set for the task type AC: "If the Probability of the Task Type (PTT) AC is Low (pertinence 85 %) and the Hit Rate of Task Type (TTT) is High (63 % pertinence) then the Need for Task Type (DTT) AC is Moderate (pertinence 74 %)", as shown in Fig. 16.

Decision-making. Each rule activation is analyzed separately and then the linguistic value of fuzzy set assigned as Highest is chosen as the best option for each feature of the new task that is being generated.

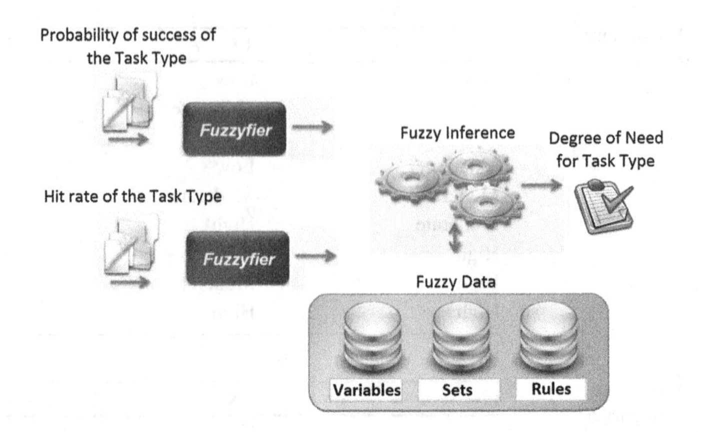

Fig. 16 Application of fuzzy logic to the task type

The logic presented above is applied to establish Type of Task, the number of incorrect comparisons and choice of words. For the words choices, instead of only one value, the algorithm that chooses the words returns a list with *n* words, where *n* is previously defined by the fuzzy logic responsible for deciding the number of comparisons.

There is no need, in this work, to defuzzify the output variables because the choice is given by the highest degree of pertinence, as previously stated. This leads to the generation of task features for the subsequent level of the game. It should be decided which task type is better to choose.

6 Results

The work is still in its initial phase and to validate the operation and efficiency of the proposal, the Fuzzy System algorithms were implemented.

The ALEPP and GEIC teaching programs are currently being tested at schools in the city of So Carlos, SP—Brazil. This database, which has records of the teaching program for each child, was used to yield the ML results. It generated data for the initial tests in the fuzzy system. These tests were conducted under supervision of specialists, as well as on the basis of a results analysis.

In order to evaluate the artificial intelligence in the game and collect some preliminary results, simulations were performed on three groups of students: (i) Students with Learning Deficit (DAP), (ii) with Gradual Learning (APG) and (iii) Students with Consolidated Learning (APC).

The system was tested in a teaching unit containing fifteen words used by GEIC and with the same teaching structure. The words were: bolo (cake), tatu (armadillo), vaca (cow), bico (beak), mala (suitcase), tubo (tube), pipa (kite), cavalo

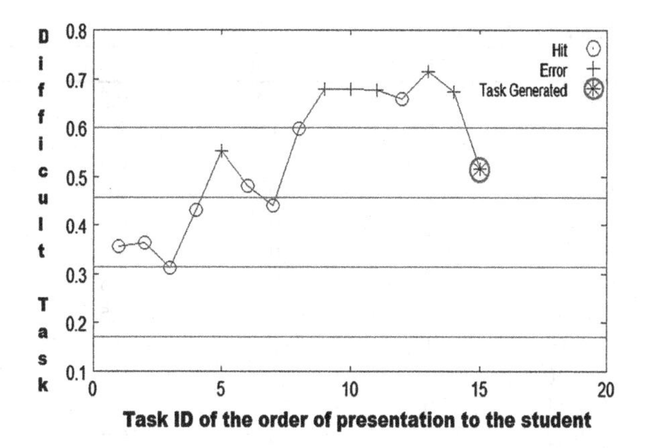

Fig. 17 Graphs with the teaching tasks of teaching the word "bolo" in the Learning Deficit group

(horse), apito (whistle), luva (gloves), tomate (tomato), vov (grandpa), muleta (crutch), fita (tape) and pato (duck).

Of the fifteen words that correspond to the teaching session called pre-test, five were analyzed regarding literacy. These five words were: bolo (cake), tatu (armadillo), apito (whistle), tomate (tomato) and muleta (crutch). The fifteen adaptative tasks generated by the fuzzy system were also analyzed by experts.

The simulations generated graphs for every single word all three behaviors. As an example, there is a graph in Fig. 17 with the tasks used for teaching the word "bolo", both for writing and for reading in the experiment with students with learning deficit. Each task has a representation of right or wrong, as indicated. Also in Fig. 17, a task with decreased difficulty is displayed, represented by a "*", which was generated by the AI based on historical errors of this student.

To validate the data generated by the AI, questionnaires were submitted to a group of psychologists containing a sequence of teaching tasks for every word of the experiment for each simulated behavior. The questionnaires had two goals. The first objective was to identify the difficulty level for each teaching task in a scale (very easy = 1, easy = 2, regular = 3, difficult = 4, very difficult = 5). The second was to, through five tasks as alternatives, check whether the psychologists should be able to answer the question: "What task the student should do based on previous tasks?", where one of the alternatives generated by the task was the AI. An average level of difficulty was calculated taking into consideration the answers given by psychologists, as seen in the following Tables 17 and 18.

The result generated by the AI and the result of the questionnaire was compared and is shown in Table 17. The lines correspond to the tasks presented to the psychologists, and the columns represent the word taught by the tasks. The table values mean the difference between the difficulty level of the task chosen by the psychologist in relation to the difficulty level of the task generated by the AI. A value near 0 indicates that the tasks generated by the AI and by the psychologists are the same, while a value near 5 indicates that the tasks are very different. For example, task 1 (row 1) that teaches the word "bolo" (column 1) and has a 3.5 difficulty level rated by AI. Psychologists, through the survey, defined the task

Table 17 Activations referring to incorrect words

	Bolo	Tatu	Apito	Tomate	Muleta
Task 1	1.5	1.5	0.5	2.5	1.5
Task 2	0.5	1.5	2	1	2
Task 3	0.5	0.5	0.5	0.5	0.5
Task 4	1	0.5	0.5	1	0.5
Task 5	1	0	0	1	0
Task 6	1	0.5	0.5	1.5	0.5
Task 7	1	0.5	0.5	0	0.5
Task 8	4	4	2.5	3.5	3.5
Task 9	1	0.5	0.5	1	1
Task 10	1	1	0.5	0.5	0.5
Task 11	0	0	0	1	1
Task 12	2	0.5	0.5	1	0.5
Task 13	0	0.5	1.5	0.5	0.5
Task 14	1	0.5	0.5	0	0

difficulty level as 2.0. The difference between the value generated by AI and the value selected by psychologists was equal to 1.5.

According to Table 17, it can be seen that:

- Tasks considered complex by the AI were also considered complex by the psychologists.
- In all the words on task 8, the difference in the difficulty level was equivalent or near 4. This is because these tasks are "Copy" type task, where the model is a text and the choices are syllables (CE type). As an improvement for this task type, it should be advised to set the score for that type and fit it to the opinions of experts.
- 14.28 % of the tasks were classified equally by the psychologists and AI, with the same level of difficulty, meaning that the difficulty level selected by the psychologist was the same as the system-generated difficulty.
- 64.28 % of the tasks have been classified with a difference between 0 and 1 point (rounding up or down), which means that a small difference is acceptable between the choice of the AI and the psychologists for the task.
- 78.56 % of the tasks are similar and acceptable by the psychologists and the AI.

Table 18 shows the data collected with the opinions of the psychologists about the difficulty level of the chosen task as appropriate for the student based on tasks

Table 18 Activations referring to incorrect words

	DAP		APG		APC	
	AI	Psychologists	AI	Psychologists	AI	Psychologists
Bolo	4	2.5	4	3.5	5	5
Tatu	4	2	4	4.5	5	5
Apito	5	2.5	5	3.5	4	3.5
Muleta	2	2	5	4.5	4	4.5

previously performed and the difficulty level generated by the AI. The data was arranged by groups of students obtained from the simulation that had been proposed. The average difficulty levels for the chosen tasks defined by the psychologists and stated in Table 18 were also calculated. The closest the value generated by the AI to the value given by the Psychologists, the better the result.

According to data collected with this simulation it was concluded that the difficulty of the tasks generated by the AI are close to the options chosen by experts.

Experts analyzed and chose one among five tasks, one that best meets the needs of the student. Regarding the tasks generated by the AI, which were hidden among the others, 25 % were considered optimal, 41.66 % were considered satisfactory and 33.34 % were considered far from ideal.

7 Conclusion and Future Studies

This work is part of a project that proposes the use of Intelligent Agents attached to computer games and targets learning of reading and writing. This part of the system being researched aims at exploring the question of decision-making when there is uncertainty about the real need for a particular person to perform a given task type.

The first results were obtained by processing data from the database project where there is a record of how the GEIC program was implemented. With this information it was possible to simulate results to allow the ML to convey on information to the Fuzzy System, and then it was validated in a satisfactory way by specialists in the field of psychology. However, it is still necessary to make adjustments to the variables, address other tasks, examine other fuzzy variables and also cover the tests directly in the computer games with the students running the system in real time.

It is concluded that the project is viable, with positive results encouraging us to continue expanding the research to enable the game to teaching writing and seeking improvements for it.

Despite obtaining satisfactory results, there is still a need to carry out tests in a complete game running in real time. It is also expected that there can be an extension of the analysis by the ML. At the same time we must include AIED in the game that is under development. The game that is currently under development is being implemented in a Game Engine called Unity3D. This tool allows various games to be generated for platforms such as PC, Xbox360, Web HTML, Flash, iOS and Android. The goal is to ensure that the end of the game runs on a tablet with Android.

References

1. Xiangfeng, L., Xiao, W., Jun, Z.: Guided Game-Based Learning. IEEE Transactions on Learning Technologies, USA (2010)
2. Benedict, B.: What does the AI in AIED buy? IEE, Savoy Place, London WC2R 0BL, U.K. (1997)
3. Tobias, S.: Interest, prior knowledge, and learning. Rev. Educ. Res. **64**(1), 3754 (1994)
4. Brophy, J.: Teacher influences on student achievement. Am. Psychol. **64**(1), 3754 (1986)
5. Dormans, D., Sander, B.: Generating Missions and Spaces for Adaptable Play Experiences. IEEE Transactions on Computational Intelligence AI in Games, USA (2011)
6. Marques, L.B., Meio, R.G., Maria, M.R.: Manual do Usurio de Programas de Ensino via GEIC—Volume 1: Aprendendo a Ler e Escrever em Pequenos Passos. So Carlos (2011)
7. Azevedo, M.A., Marques, M.L.: Alfabetizao hoje. Cortez, So Paulo (2001)
8. Sarmanho, E.S., Sales, E.B., Cavalcante, D.M., Marques, L.B.: Um Jogo com Reconhe-cedor de Voz para o Ensino de Crianas com Dificuldade de Aprendizagem em Leitura e Escrita. Semish (2011)
9. Schaefer, E., Brevik, M., Sexton, E.: Diablo. Blizzard North (1996)
10. Baldree, T.: Torchlight. Runic Games, England (2009)
11. Wright, W.: Spore. Maxis (2008)
12. Person, M., Bergensten, J.: Mojang, Sweden (2009)
13. Rose, J.C., Souza, D.G., Rossito, A.L., Rose, T.M.S.: Aquisio de leitura aps histria de fracasso escolar: equivalncia de estmulos e generalizao. In: Psicologia: Teoria e Pesquisa. 45169 (1989)
14. Souza, D.G., Rose, J.C., Faleiros, T.C., Bortoloti, R., Hanna, E.S.: Teaching generative reading via recombination of minimal textual units: a legacy of verbal behavior to children in Brazil. In: 19th International Journal of Psychology and Psychological Therapy, pp. 19–44 (2009)
15. Reis, T.S., Souza, D.G., Rose, J.C., Avaliao de um programa para o ensino de leitura e escrita. In: Estudos em Avaliao Educacional, vol. 20, pp. 425–450 (2009)
16. Nerino, G.Jr., Pereira, A.B.P., Cavalcante, D.M., Sales, E.B., Marques, L.B.: Mquina de aprendizagem como ferramenta de auxlio na anlise comportamental no ensino da leitura. In: Renote- Revista sobre novas tecnologias na Educao, vol. 10 (2012)
17. Pedro, C.D., Adriano, J.O.: Aprendizado de Regras Nebulosas em Tempo Real para Jo-gos Eletrnicos. XI Brazilian Symposium of Multimedia Systems and Web. Games II Brazilian Workshop of Games and Digital Entertainment (2003)
18. Moratori, P.B., Pedro, M.V., Manhaes, L.M.B., Lima, C., Cruz, A.J.O., Ferreira, E.B., Andrade, L.C.V.: Analysis of the stability of a Fuzzy control system developed to control a simulated robot. In: 14th IEEE International Conference on Fuzzy, pp. 726–730. IEEE Press, New York (2005)
19. Marques, L.B.: Variveis Motivacionais no Ensino de Leitura: O jogo como recurso complementar. UFSCar, So Carlos (2009)
20. Mamdani, E.H.: Advances in the linguistic synthesis of fuzzy controllers. Int. J. Man Mach. Stud. **8**, 669–678 (1976)
21. Timothy, R.J.: Fuzzy Logic with Engineering Applications, vol. 5, pp. 117–148. ISBN: 047074376X (2010)

Optimizing the Performance of a Refrigeration System Using an Invasive Weed Optimization Algorithm

Roozbeh Razavi-Far, Vasile Palade and Jun Sun

Abstract This paper presents a study and the obtained results on the performance optimization of a large refrigeration system in steady state conditions. It is shown that, by using adequate knowledge on plant operation, the plant wide performance can be optimized with respect to a small set of variables. For this purpose, an appropriate performance function is defined. A derivative free optimization technique based on the invasive weed optimization (IWO) algorithm has been used to optimize the parameters of the local controllers in the system. The performance of the IWO algorithm, both in terms of optimality of the results and speed of convergence, is compared with particle swarm optimization (PSO) algorithm. Simulation results have been used to validate the proposed approach.

Keywords Performance optimization · Refrigeration systems · Invasive weed optimization (IWO) · Particle swarm optimization (PSO)

R. Razavi-Far (✉)
Department of Control Engineering and System Analysis, Université Libre de Bruxelles (ULB), 50 Av. F.D. Roosevelt,CP 165/55 B-1050 Brussels, Belgium
e-mail: roozbeh.razavi-far@ulb.ac.beroozbeh.razavi@gmail.com

R. Razavi-Far
Danfoss A/S, A/C Control, Nordborg, Denmark

V. Palade
Department of ComputerScience, University of Oxford, Wolfson Building, Parks Road, Oxford OX1 3QD, UK
e-mail: vasile.palade@cs.ox.ac.uk

J. Sun
Department of Computer Science and Technology, Jiangnan University, No. 1800, Lihu Avenue 214122 Wuxi, Jiangsu, China

I. Hatzilygeroudis and V. Palade (eds.), *Combinations of Intelligent Methods and Applications*, Smart Innovation, Systems and Technologies 23,
DOI: 10.1007/978-3-642-36651-2_5,

1 Introduction

Complex plants, such as large supermarket systems, include many sub-systems that dynamically interact together. Plant-wide performance optimization is not guaranteed by just local tuning of individual controllers of the subsystems.

Various optimization techniques have been extensively used to tune the controller parameters to achieve an improved performance [1, 2]. However, these techniques are generally gradient-based and mostly focused on local performance optimization.

An alternative is to use derivative-free search algorithms. These algorithms directly utilize the performance function and constrain values to steer towards the optimal solution. Recently, genetic algorithms [3], particle swarm optimization [4], ant colony optimization [5], simulated annealing [6] and tabu search [7] have been widely used for global optimization in different engineering applications [8, 9].

The Invasive Weed Optimization (IWO) algorithm is a bio-inspired numerical optimization algorithm that simulates the behavior of weeds in nature when colonizing and finding a suitable place for growth and reproduction [10]. Here, the IWO algorithm is used to find the optimal parameters for the local controller with respect to a plant-wide performance function. A comparative evaluation of the IWO algorithm and particle swarm optimization (PSO) algorithm, on this problem, is carried out in order to validate the appropriateness of the simulation results obtained.

The rest of the paper is organized as follow. The description of the refrigeration system under optimization is presented in Sect. 2. To perform the optimization task, an appropriate performance function is proposed in Sect. 3 along with the problem formulation. The IWO algorithm, its main properties and pseudocode are presented in Sect. 4. In Sect. 5, the main features and algorithm of a standard PSO are briefly explained. Simulation setup and comparative results are shown in Sect. 6. Conclusions are drawn in Sect. 7.

2 System Description

This section provides a short description of the vapor-compression cycle in a refrigeration system. Then, the relevant dynamics of the system, seen from a control perspective, are described. In the vapor-compression cycle a refrigerant is circulating between two heat exchangers. One in the cold storage where the refrigerant absorbs energy by evaporating, and one in a hot reservoir, typically the surroundings, where energy leaves the refrigerant during condensing [11].

The temperature at which the refrigerant evaporates / condenses is called the saturation temperature, T_{sat}. For the cycle of heat transfers in the refrigeration system to work, T_{sat} in the evaporator must be lower than the temperature in the cold storage, while it, in the condenser, must be higher than the temperature of the surroundings. To achieve this a compressor is inserted between the evaporator and the condenser since T_{sat} of any liquid or gas depends on the pressure. Hence, evaporator and condenser are referred to as the low and high pressure parts respectively.

Figure 1 shows the layout of the refrigeration system with four states of the process marked in the figure and Fig. 2 shows the correspondence between pressure and specific enthalpy in the refrigerant for the entire process with markings of the same four states. Both figures are from [11]. When the refrigerant enters the evaporator as a liquid it starts evaporating. Thus, the first section of the evaporator contains a mixture of liquid and gas and is called the two-phase region. In order not to damage the compressor, it is very important that no liquid flows out of the evaporator outlet, and for that reason, an important control objective is to make sure that the two-phase region ends before the evaporator outlet such that the last section of the evaporator only contains gas. This section is called the superheat region. In the followings, we provide the model of relevant parts that are used for control purposes. The following references [12–15] have been used, to derive the dynamic equations for the model.

2.1 The Expansion Valve

The refrigerant mass flow through the valve can be modeled as [15]:

$$\dot{m}_i = OD\alpha\sqrt{P_c - P_e} \tag{1}$$

In the above equation, $\dot{m}_i$ refrigerant mass flow rate at evaporator inlet, P_c and P_e are condensing and evaporating pressure respectively, α is the heat transfer coefficient, and OD is the opening degree of the expansion valve ($0 \leq OD \leq 1$).

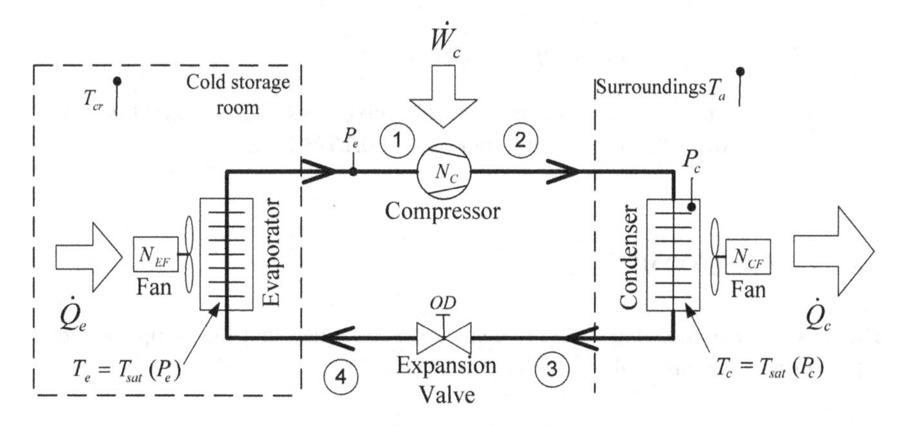

Fig. 1 The basic layout of a refrigeration system [11]

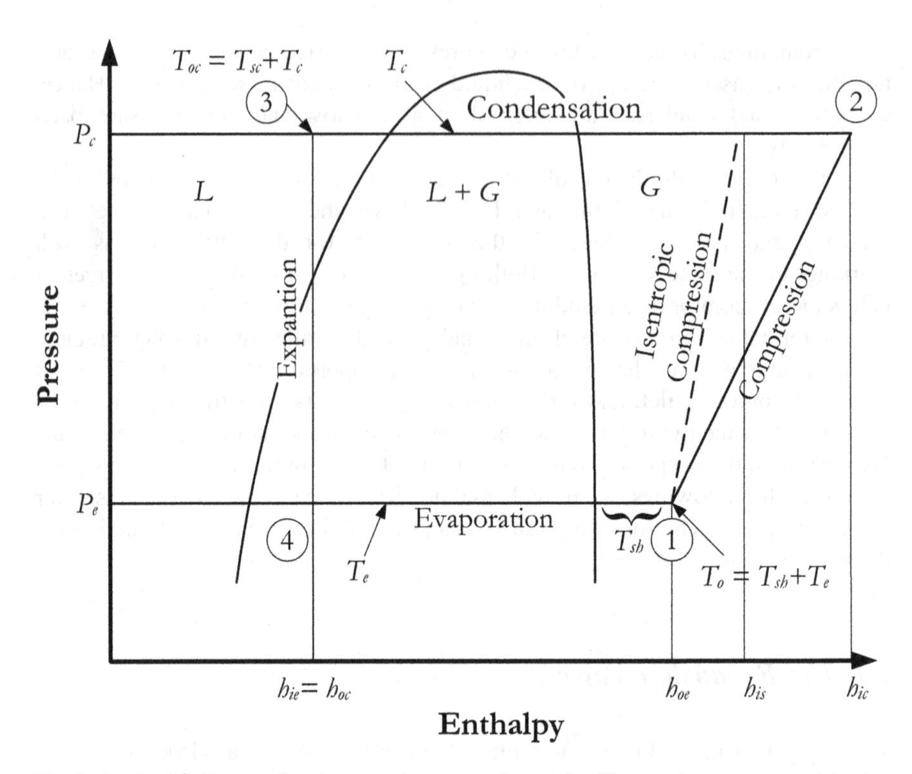

Fig. 2 h - log(p)-diagram for the vapour-compression cycle. The numbers correspond to the states in Fig. 1. Subindices : i = inlet, o = outlet, e = evaporator, c = condenser. L and G denotes Liquid and Gas respectively [11]

The mass flow $\dot{m}(kg/s)$ characteristics can be experimentally determined as a function of opening degree OD, and presented by the following polynomial function [15]:

$$\begin{aligned} \dot{m}_i &= f_{OD}(OD) \\ &= a_1 + a_2OD + a_3OD^2 + a_4OD^3 + a_5OD^4 \end{aligned} \tag{2}$$

For the chosen case the parameters are given as $a_1 = -0.009689, a_2 = 0.2236, a_3 = -0.4178$, $a_4 = 0.4546$, and $a_5 = -0.1573$ [15].

2.2 The Compressor

The mass flow is modeled in Eq. (3), since it is assumed that the compressor can act as an ideal pump, with the volume of V_{comp} [13].

$$\dot{m}_o = \rho_g\ V_{comp} f_{comp}, \tag{3}$$

where $\dot{m}_o$ stands for refrigerant mass flow rate at evaporator outlet, f_{comp} denotes the compressor speed (in frequency), and ρ_g is the density of the refrigerant in gas form. The suction pressure P_e is considered to be proportional to the vapor density in all operating conditions [13]. Therefore, the mass flow in Eq. (3) can be reformulated by means of a constant α_{comp} as:

$$\dot{m}_o = \alpha_{comp} P_e f_{comp} \tag{4}$$

2.3 The Evaporator Dynamics

For the design and optimization of controller parameters, an elaborate description of the evaporator dynamics is not needed since, in real life, such detailed models do not exist. To attain a proper model, the control signal *OD* is mapped to the superheat temperature *Sh*. This I/O mapping, is done by carrying out an *OD* sweep on the refrigeration system while keeping the air temperature $T_{air,in}$, the compressor frequency f_{comp}, and the condenser pressure P_c constant.

The dynamic behaviors of the air temperature T_a, outlet temperature T_o, and evaporation temperature T_e with respect to a gradual increase in opening degree *OD* of the expansion valve are presented in Fig. 3 [15]. Earlier studies, in [15],

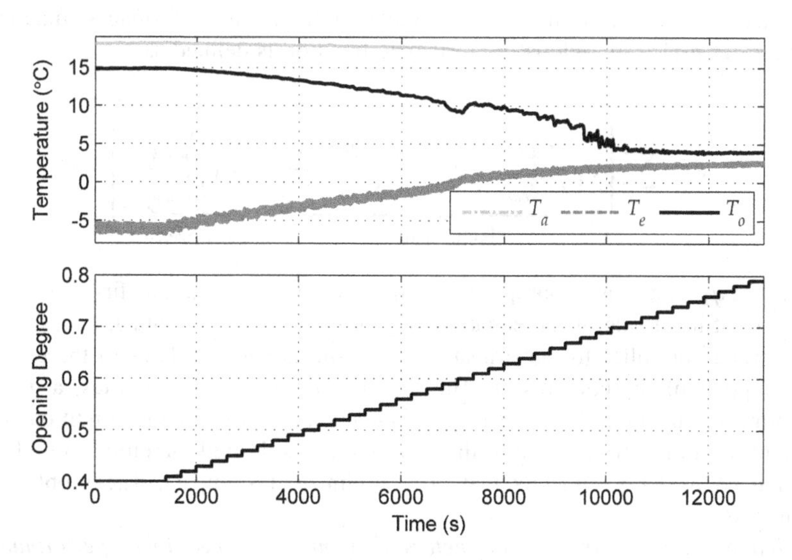

Fig. 3 The temperature profiles as a function of a sweep on the expansion valves opening degree (OD). The temperatures are air temperature T_a, Outlet temperature T_o, and evaporation temperature T_e [15]

revealed that there is a considerable change in the system gain. This system nonlinearity is appropriately described, in [16], by the inverse trigonometric function *atan* expressed as:

$$y = 7\left(-atan\left(10\pi\frac{u-50}{50}\right)+\frac{\pi}{2}\right) \tag{5}$$

In the above equation, u and y stand for the input and output respectively, and the rest is arbitrary scaling [16]. In addition, the transfer function of the outlet temperature T_o to opening degree OD, in the Laplace domain, is explained by a first-order-plus-dead-time (FOPDT) process model [15]:

$$H(s) = \frac{K_p e^{-Ls}}{\tau s + 1}, \tag{6}$$

where K_p stands for the lumped system gain, L denotes the time delay, and τ is the system time constant. The transfer function parameters vary with respect to the change in operating points [15].

3 Problem Formulation

The dynamic behavior of the evaporator unit varies due to the change in the operating conditions at the compressor unit. Therefore, it would be interesting to investigate whether it is possible to optimize the parameters of the evaporator's superheat controller, from the perspective of a global performance function. Inspired from [15], the global performance function is defined as:

$$J = \sum_{k=0}^{N}\left(\underbrace{q_1\frac{OD_k^2}{OD_{max}^2} + q_2\frac{(Sh_k - Sh_{ref})^2}{Sh_{ref}^2}}_{\text{Evaporator}} + \underbrace{q_3\frac{f_{comp,k}^2}{f_{compo}^2}}_{\text{Compressor}}\right) \tag{7}$$

where q_1, q_2, and q_3 are appropriate positive weight constants. The first two terms in the right-hand side of the equation are normally used for evaluation/design of appropriate controller for superheat control. The last term relates to the energy consumption of the compressor. The performance function J is, thus, a global performance function. Since a change in the compressor operating conditions has an impact on the performance of the controller, it is desired therefore, to find the optimal controller parameters that also minimize this impact. The problem is defined as:

Find a set of controller parameters that minimize the global performance function J over all operating conditions.

Using *derivative-based* optimization methods can become problematic due to the inherent non-smoothness in the performance function, which can make the computation of finite differences very inaccurate. The non-smoothness is caused by discrete switchings in the compressor speed.

4 Invasive Weed Optimization

At the beginning, before explaining the IWO algorithm, the key terms [10, 17] are introduced here as follow:

Seed: each individual in the colony that includes a value for each variable in the optimization problem prior to fitness evaluation.

Fitness: a value represents the merit of the solution for each seed.

Weed/Plant: each evaluated seed grows to a flowering plant or weed in the colony. Therefore, growing a seed to a plant corresponds to evaluating an individual's fitness.

Colony: the search space and indicates all agents or seeds.

Population size: the number of plants in the colony.

Maximum weed population: a predefined parameter corresponds to the maximum allowed number of weeds in the colony posterior to fitness evaluation.

The following steps are considered to simulate the colonizing behaviour of weeds [10, 17]:

1 **Search space definition**: Initially, the number of parameters that need to be optimized has to be defined, hereafter denoted by D. Next, for each parameter in the D-dimensional search space, a minimum and maximum value are assigned.
2 **Population initialization**: A limited number N_0 of initial seeds are being randomly dispread through the defined search space, $W = \{w_1, w_2, \ldots, w_{N_0}\}^T$. Consequently, each seed catches a random position in the D-dimensional search space [10].
3 **Fitness estimation**: A fitness value assigned to each initial seed by the fitness function, defined to represent the goodness of the solution [17]. Here, initial seeds grow up to flowering plants.
4 **Ranking and reproduction**: The flowering plants are firstly ranked based on their assigned fitness values with respect to others. Subsequently, flowering plants can reproduce new seeds with respect to their rank in the colony. In other words, the number of seeds produced by each plant can increase linearly from the minimum possible seeds production S_{min}, to its maximum, S_{max} based on their own, the lowest, and the highest fitness of the colony (all plants). Then, the plants with higher fitness which are more adapted to the colony can produce more seeds that solve the problem better [17]. Figure 4 illustrates this procedure [10]. The number of seeds to be created by each plant is computed as follow:

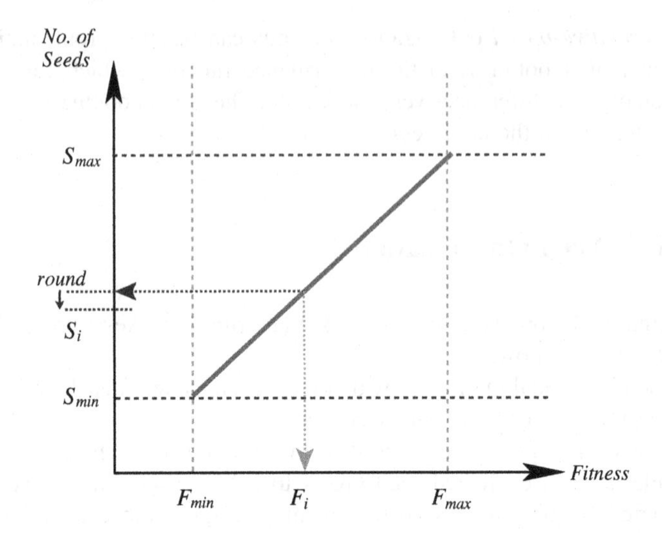

Fig. 4 Seed reproduction procedure [10]

$$S_i = \left\lfloor \frac{F_i - F_{min}}{F_{max} - F_{min}} (S_{max} - S_{min}) + S_{min} \right\rfloor \tag{8}$$

where F_i is the fitness of $i-th$ plant. F_{min} and F_{max} stand for the lowest and highest fitness in the weed population. Therefore, this step guarantees cooperation of every weed in the reproduction procedure. Note that S_{max} and S_{min} are predefined parameters of the algorithm and adjusted according to the structure of the problem.

5 **Spatial dispersal**: Randomness and adaptation is provided in this part of the algorithm [10]. Here, the seeds are being randomly scattered through the search space by using normally distributed numbers with zero mean and adaptive standard deviations [10] as follow:

$$w_s[\kappa] = w[\kappa] + \mathcal{N}(0, \sigma_{iter}^2) \tag{9}$$

where, $w[\kappa]$ indicates the $\kappa - th$ variable of a solution vector in the current iteration, and $w_s[\kappa]$ shows the $\kappa - th$ variable of its $s - th$ seeds. The standard deviation σ_{iter} at the present time step can be computed adaptively according to the following equation [10]:

$$\sigma_{iter} = \frac{(iter_{max} - iter)^n}{(iter_{max})^n} \left(\sigma_{initial} - \sigma_{final}\right) + \sigma_{final} \tag{10}$$

where $\sigma_{initial}$ and σ_{final} denote to the pre-defined initial and final standard deviations, respectively. $iter_{max}$ indicates the maximum allowed number of iteration cycles and n is the nonlinear modulation index [10] assigned by the user. The

σ_{iter} can be reduced from the $\sigma_{initial}$ to the σ_{final} with different velocities in accordance with the chosen nonlinear modulation index, n.
Initially, the whole search space can be explored by the algorithm due to the high value of initial standard deviation $\sigma_{initial}$. Then, the standard deviation σ_{iter} is gradually reduced by increasing the number of iterations, to focus the search around the local minima or maxima to find the global optimum. This gradual reduction guarantees to collect only fitter plants and to discard plants with lower fitness. The produced seeds, along with their parents are considered as the potential solutions for the next population.

6 **Competitive exclusion**: After passing a number of iterations, the population size reaches its pre-defined maximum (P_{max}) by fast reproduction and consequently a mechanism for discarding the plant with low fitness will be activated. To this end, the seeds and their parents are ranked together and those with higher fitness survive and subsequently reproduce new seeds in the next iteration.

7 **Termination condition**: Survived plants reproduce new seeds with respect to their fitness rank in the colony. The procedure is repeated at step 3 until either the maximum allowed number of iterations has been reached or the fitness criterion met [17].

A pseudocode version of the IWO algorithm is given in Fig. 5 [18].

5 Particle Swarm Optimization

PSO is an evolutionary search process for stochastic optimization based on the social learning metaphor [4]. The PSO algorithm mimics the social behavior in flocks of birds when they are flying, by simulating a population of potential solutions or particles 'so-called swarms' in a multidimensional search space [19]. These randomly initialized particles freely fly across the predetermined search space and update their own position and velocity according to their best experience, the best experience of the entire group 'population' and balancing exploration and exploiting [20].

The aim of the PSO algorithm is to find a set of particles 'solutions' that minimize an objective function J. In the PSO algorithm, the so-called swarm includes a set of particles $\mathcal{P} = \{p_1, p_2, \ldots, p_{SW}\}$, where SW stands for the swarm size. Each particle's position stands for a candidate solution to minimize the fitness function J. At each time step t (iteration index in the optimization context), particle p_l contains a position $\vec{x}_l^t$ and a velocity $\vec{v}_l^t$ associated to it.

The PSO algorithm is begun by initialization on which random positions are generated for the particles, within an initialization region. The velocities are initialized also to zero or to small random values to preserve the particles inside the predetermined search space during the first iterations [21].

The velocities and positions of the particles are iteratively updated based on the following movement Eqs. (11,12), until a stopping criterion has been met [21].

INPUTS: N_0, S_{min}, S_{max}, $iter_{max}$, $\sigma_{initial}$, σ_{final}, n, P_{max} and J
GENERATE a random population of N_0 individuals from a set of feasible solutions

$$W = \{w_1, w_2, \ldots, w_{N_0}\}^T$$

DO FOR $iter = 1, 2, \ldots, iter_{max}$
EVALUATE the fitness function for each individual in W.
COMPUTE the maximum and minimum fitness in the colony F_{max} and F_{min}.
DO FOR each individual w_i
COMPUTE number of seeds for w_i

$$S_i = \lfloor (F_i - F_{min})\,(S_{max} - S_{min}) \,/\, (F_{max} - F_{min}) + S_{min} \rfloor$$

RANDOMLY distribute seeds over the search space with normal distribution $\mathcal{N}\left(0, \sigma_{iter}^2\right)$ around the parent w, with 0 mean & an adaptive standard deviation

$$\sigma_{iter} = \sigma_{final} + (\sigma_{initial} - \sigma_{final})\,(iter_{max} - iter)^n \,/ (iter_{max})^n$$

ADD the generated seeds to the solution set, W.
END DO
IF $(|W| = N) > P_{max}$, Then:
SORT the population W in descending order of their fitness.
TRUNCATE population of weeds with smaller fitness until $N = P_{max}$;
END IF
END DO
BEST solution is the plant w_{best} with minimum fitness in the last population.

Fig. 5 The pseudo-code for the IWO algorithm [18]

$$\vec{v}_l^{t+1} = \psi^t \vec{v}_l^t + c_1 \vec{\varphi}_1^t (\vec{b}_l^t - \vec{x}_l^t) + c_2 \vec{\varphi}_2^t (\vec{g}_l^t - \vec{x}_l^t) \tag{11}$$

$$\vec{x}_l^{t+1} = \vec{x}_l^t + \vec{v}_l^{t+1} \tag{12}$$

where $\vec{b}_l^t$ is the best position of the lth particle (with respect to J) discovered thus far, $\vec{g}_l^t$ is the global best position of the swarm until t^{th} iteration, and l stands for the particle's index, $l = 1, \ldots, SW$. c_1 and c_2 are the acceleration factors. φ_1^t and φ_2^t are diagonal matrices in which the main diagonal entries are uniformly distributed random numbers in the range $[0, 1)$ and iteratively updated at each iteration. ψ^t is the time-varying inertia weight [22]. The inertia weight linearly decreases in each iteration, based on the following:

$$\psi^t = (\psi_{max} - \psi_{min}) \frac{t_{max} - t}{t_{max}} + \psi_{min} \tag{13}$$

where ψ_{max} and ψ_{min} are the maximum and minimum values of the inertia weight, respectively, and t_{max} is the maximum allowed number of iterations. The particles move in the search process based on their own best position and the global best position, in a cooperative manner, until best results can be found [21]. Here, all particles of the swarm are fully connected together, and thus, all particles are considered as neighbors. A pseudocode version of the PSO algorithm is given in Fig. 6 [21].

6 Simulation Setup and Results

In this section, the invasive weed optimization algorithm is used for plant-wide optimization of the supermarket refrigeration system and compared with PSO. The simulation setup and results are presented in this section.

6.1 Simulation Setup

The optimization approach has been tested on a simulation model of a supermarket refrigeration system. The simulation setup is presented in Fig. 7.

The IWO algorithm has been applied to optimize the integration time, T_n, and the gain, K_p, of the PI controller. Some constraints are needed for T_n and K_p, since

```
INPUTS: J, t_max, ψ_max, ψ_min, c1, c2 and SW
SET t = 0
DO FOR l = 1, 2, ..., SW
   INITIALIZE random particle's position x_l^t
   INITIALIZE random particles' velocity v_l^t
   CALCULATE fitness value for each particle
   INITIALIZE the local best b_l^t
   INITIALIZE the global best g_l^t
END DO
WHILE t = 1, 2, ..., t_max
   DO FOR l = 1, 2, ..., SW
      GENERATE random matrices φ_1^t, φ_2^t within [0, 1)
      UPDATE inertia weight ψ^t according to Eq.(13)
      UPDATE particles' velocity v_l^t according to Eq.(11)
      UPDATE particle's position x_l^t according to Eq.(12)
      UPDATE the fitness value of each particle
      UPDATE particle's local best b_l^t
      UPDATE swarm's global best g_l^t
   END DO
SET t = t + 1
END WHILE
OUTPUT: Best solution found.
```

Fig. 6 The pseudo-code for the PSO algorithm [21]

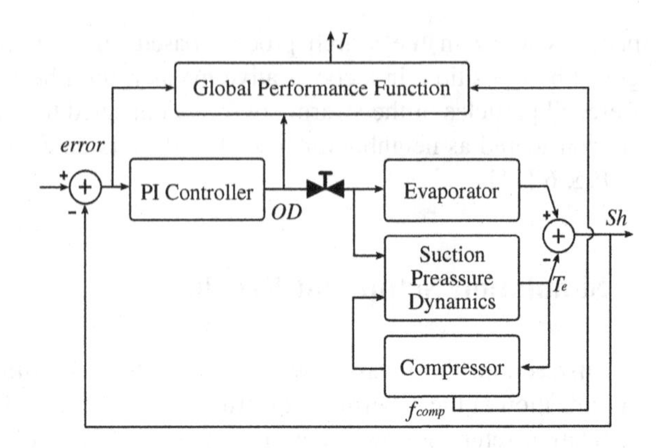

Fig. 7 Simplified simulation setup

Table 1 The IWO and PSO parameters

Quantity	Symbol	Value
Number of initial plants	N_0	5
Maximum number of iterations	$iter_{max}$	200
Problem dimension	D	2
Maximum number of plants	P_{max}	30
Maximum number of seeds	S_{max}	5
Minimum number of seeds	S_{min}	0
Nonlinear modulation index	n	3
Initial value of standard deviation	$\sigma_{initial}$	1
Final value of standard deviation	σ_{final}	0.05
Acceleration factors	c_1, c_2	2
Maximum inertia weight	ψ_{max}	0.9
Minimum inertia weight	ψ_{min}	0.4
Swarm size	SW	30
Maximum number of PSO's iterations	t_{max}	200

the feasible set for each variable can not overlap together. These constrains have been applied to the search space according to the knowledge about the controller behavior.

The key parameters of the IWO algorithm (e.g. $\sigma_{initial}$, σ_{final} and n) have a crucial affect on its own convergence. The careful tuning of these parameters can guarantee having a proper value of the standard deviation at each iteration, σ_{iter}. The maximum number of plants in the colony P_{max} should also be selected a priori based on the achievable performance of the algorithm. The key parameters of concern and their selected values to start the IWO simulation are listed in Table 1.

The maximum number of simulation runs by the IWO algorithm, $SR_{max} = N_0 \cdot (S_{max} + 1) + P_{max} \cdot (iter_{max} - 2)$ equals to 5970. The simulation is performed for 6000 seconds, to guarantee a steady state performance function.

6.2 Results

In Fig. 8, the minimum fitness value J_{min} or the best performance of the refrigeration system is plotted for each iteration of the IWO algorithm. The optimal values proposed by the IWO algorithm for the controller parameters after 200 iterations, $T_n = 58.7295$ and $K_p = -0.1536$, lead to a reasonable performance, 1439466.

To evaluate the performance of the IWO algorithm, the simulation results are compared with the results of the PSO algorithm, introduced in Fig. 6. The key parameters of the PSO algorithm for the best results used for simulation are listed in Table 1.

The PSO algorithm produces the results, J_{min}, shown on Fig. 8 after 200 iterations. The use of initial values for the algorithms can decrease the number of iterations and consequently reduce the convergence time to find the optimal solution, however the parameters of the both algorithms are properly chosen to provide an identical setup for a fair comparison. The optimal values proposed by the PSO algorithm for the controller parameters after 200 iterations, $T_n = 61.7482$ and $K_p = -0.1465$, lead to a performance, 1547500. The results of IWO and PSO algorithms are compared in Table 2 by means of the minimum fitness value J_{min}, maximum fitness value J_{max}, mean fitness value J_{mean} and standard deviation J_σ of the population in the last iteration of the simulation.

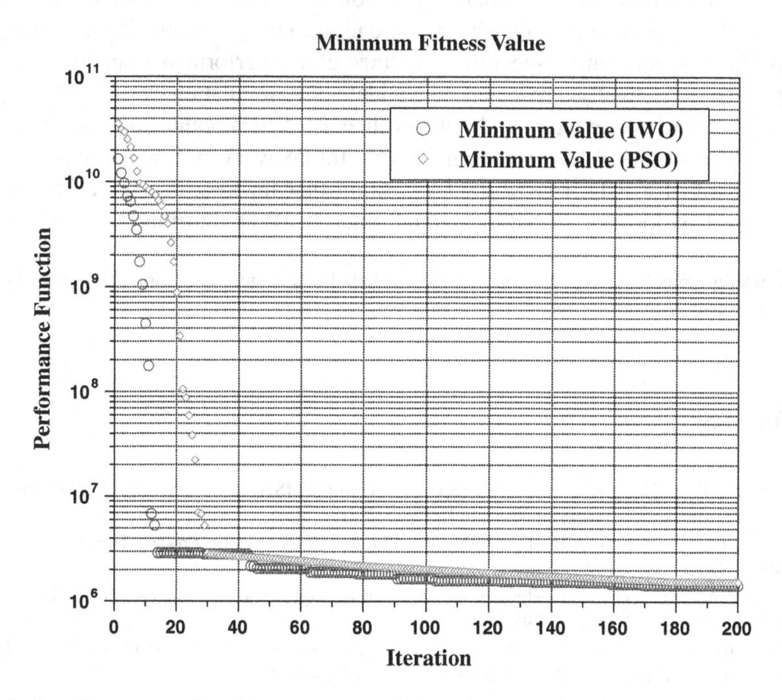

Fig. 8 Iterations versus the minimum values of the performance function J_{min}

Table 2 Comparison of J_{min}, J_{max}, J_{mean} and J_{σ} of the population in the last iteration of IWO and PSO simulations

	J_{min}	J_{max}	J_{mean}	J_{σ}
IWO	1439466	1596000	1475000	89563
PSO	1547500	1706716	1634000	103693

As shown, in Table 2 and Fig. 8, the IWO algorithm outperforms the PSO. J_{min} is dramatically decreased in few iterations by both the IWO and the PSO algorithms, and both algorithms converge rapidly, however the IWO has slightly better results than PSO (Table 2). Here, the comparison was done in order to validate the results obtained by the IWO algorithm in this particular control problem are sensible. However, both algorithms and their hybrid schemes can be used in many other optimization problems from engineering, and not only.

7 Conclusion

This paper focused on the application of the IWO and PSO algorithms for the optimization of the global system performance of a supermarket refrigeration system. An appropriate performance function was firstly introduced. Then, the IWO and PSO optimization algorithms were used to find the optimal parameters of the local controllers with respect to the defined global performance function. A fair comparison was finally presented to validate the simulation results. A proper choice of the key parameters and initial values is of paramount importance in the performance of the IWO algorithm. In our future work, we will focus on the development of a hybrid IWO-PSO algorithm that incorporates some PSO concepts, like best local or global individual, into the IWO algorithm.

Acknowledgments The authors thank Dr. Roozbeh Izadi-Zamanabadi, for support in sstem description and problem formulation.

References

1. Aström, K.J., Hägglund, T.: Advanced PID Control. ISA—Instrumentation Systems and Automation Society, Research Triangle Park (2006)
2. Smith, C.A., Corripio, A.B.: Principles and Practice of Automatic Process Control, 2nd edn. Wiley, New York (1997)
3. Goldberg, D.E.: Genetic Algorithms in Search, Optimization, and Machine Learning. Addison Wesley, Reading (1989)
4. Kennedy, J., Eberhart, R.: Particle swarm optimization. In: Proceedings of the IEEE International Conference Neural Networks, pp. 1942–1948 (1995)

5. Dorigo, M., Maniezzo, V., Colorni, A.: Ant system: optimization by a colony of cooperating agents. IEEE Trans. Syst. Man Cybern. **26**, 1–3 (1996)
6. Otten, R.H.J.M., Van-Ginnekent, L.P.P.P.: The Annealing Algorithm. Kluwer Academic, Boston (1989)
7. Pham, D.T., Karaboga, D.: Intelligent Optimization Techniques. Springer, London (2000)
8. Boeringer, D.W., Werner, D.H.: Particle swarm optimization versus genetic algorithms for phased array synthesis. IEEE Trans. Antennas Propag. **52**, 771–779 (2004)
9. Razavi-Far, R., Davilu, H., Palade, V., Lucas, C.: Model-based fault detection and isolation of a steam generator using neuro-fuzzy networks. Neurocomput. J. **72**, 2939–2951 (2009)
10. Mehrabian, A.R., Lucas, C.: A novel numerical optimization algorithm inspired from weed colonization. Ecol. Inform. **1**, 355–366 (2006)
11. Larsen, L.F.S.: Model based control of refrigeration systems. Ph.D. dissertation, Aalborg University (2005)
12. He, X.-D., Liu, S., Asada, H.H., Itoh, H.: Multivariable control of vapor compression systems. HVAC &R Res. **4**(3), 205–230 (1998)
13. Rasmussen, H., Thybo, C., Larsen, L.F.S.: Nonlinear superheat and evaporation temperature control of a refrigeration plant. In: The IFAC Workshop on Energy Saving Control in Plants and Buildings (2006)
14. Rasmussen, H., Larsen, L.F.S.: Non-linear and adaptive control of a refrigeration system. IET Control Theory Appl. **5**, 364–678 (2011)
15. Izadi-Zamanabadi, R., Vinther, K., Mojallali, H., Rasmussen, H., Stoustrup, J.: Evaporator unit as a benchmark for plug and play and fault tolerant control. In: 8th IFAC Safeprocess, Mexico City (2012)
16. Vinther, K., Rasmussen, H., Izadi-Zamanabadi, R., Stoustrup, J.: Single temperature sensor based evaporator filling control using excitation signal harmonics. In: Proceedings of the IEEEMulti-Conference on Systems and Control, Dubrovnik, Croatia (2012)
17. Karimkashi, S., Kishk, A.A.: Invasive weed optimization and its features in electromagnetics. IEEE Trans. Antennas Propag. **58**(4), 1269–1278 (2010)
18. Mehrabian, A.R., Yousefi-Koma, A.: Optimal positioning of piezoelectric actuators on a smart fin using bio-inspired algorithms. Aerosp. Sci. Technol. **11**, 174–182 (2007)
19. Eberhart R., Shi, Y.: Comparing inertia weights and constriction factors. In: Proceedings of the Congress on Evolutionary Computing, pp. 84–89 (2000)
20. Jones, K.O.: Comparison of genetic algorithms and particle swarm optimization for fermentation feed prole determination. In: International Conference on Computer Systems and Tecnologies (2006)
21. Dorigo, M., Montes de Oca, M.A., Engelbrecht, A.: Particle swarm optimization. Scholarpedia **3**(11), 1486 (2008)
22. Shi, Y., Eberhart, R.: A modified particle swarm optimizer. In: Proceedings of IEEE International Conference on Evolutionary, Computing, pp. 69–73 (1998)

A New Cooperative Evolutionary Multi-Swarm Optimizer Algorithm Based on CUDA Architecture Applied to Engineering Optimization

Daniel Leal Souza, Otávio Noura Teixeira, Dionne Cavalcante Monteiro and Roberto Célio Limão de Oliveira

Abstract This paper presents a new Cooperative Evolutionary Multi-Swarm Optimization Algorithm (CEMSO-GPU) based on CUDA parallel architecture applied to solve engineering problems. The focus of this approach is: the use of the concept of master/slave swarm with a mechanism of data sharing; and, the parallelism method based on the paradigm of General Purpose Computing on Graphics Processing Units (GPGPU) with CUDA architecture, brought by NVIDIA corporation. All these improvements were made aiming to produce better solutions in fewer iterations of the algorithm and to improve the search for best results. The algorithm was tested for some well-known engineering problems (WBD, ATD, MWTCS, SRD-11) and the results compared to other approaches.

Keywords PSO · EPSO · GPGPU · GPU · CUDA · NVIDIA · Swarm · Cooperative · Parallel · Hetergeneus computing · Particle

D. L. Souza (✉) · O. N. Teixeira
Laboratory of Natural Computing (LCN) Area of Exact and Natural Sciences (ACET), University Centre of Pará (CESUPA), Belém, Pará, Brazil
e-mail: daniel.leal.souza@gmail.com

O. N. Teixeira
e-mail: onoura@gmail.com

D. L. Souza · D. C. Monteiro
Institute of Exact and Natural Sciences (ICEN), Laboratory of Applied Artificial Intelligence (LAAI), Federal University of Pará (UFPA), Belém, Pará, Brazil
e-mail: dionne@ufpa.br

O. N. Teixeira · R. C. L. de Oliveira
Institute of Technology (ITEC), Post-Graduate Program in Electrical Engineering (PPGEE), Federal University of Para (UFPA), Belém, Pará, Brazil
e-mail: limao@ufpa.br

I. Hatzilygeroudis and V. Palade (eds.), *Combinations of Intelligent Methods and Applications*, Smart Innovation, Systems and Technologies 23,
DOI: 10.1007/978-3-642-36651-2_6,

1 Introduction

In recent years, the evolution of computational resources made possible some significant advances in the techniques of search and optimization. Bastos Filho et al. [1] affirms that these advances have brought success in the construction of more efficient solutions, also in studies of metaheuristics, which guide the heuristics to obtain optimal solutions applied in N-dimensional search space problems. These problems vary from a wide range of practical applications, including industry and management of basic supplies such as electric energy, gas and petroleum refining, for example [2].

One of the most well known metaheuristics with large use in engineering fields is the Particle Swarm Optimization (PSO) [8]. Among its variants, Evolutionary Particle Swarm Optimization (EPSO) proposed in [3] stands out for including the set of operations of evolutionary strategies such as replication, mutation and selection applied in a PSO environment.

Another type of variation of PSO algorithm involves the concept of multiple population, consisting of a master swarm and slave swarms. Van Den Bergh and Engelbrecht [4] uses the model of search partitioning in genetic algorithms of PSO in order to reduce the deterioration of performance while dimensionality of search space increases. In general, most of these metaheuristics use computing systems with parallel and distributed architecture under an inter-communicable environment between two or more swarms. The ultimate goal is a more efficient search and the exchange of information obtained by each of them in order to compare and refine the search based on the results already obtained by the neighboring clusters.

In the matter of the improvement of hardware resources, it's possible to highlight the evolution of Graphics Processing Units (GPUs). Since 2003, the many-cores processors, strongly represented by GPU, had shown greater superiority in terms of speed, especially operations involving floating point data [5]. Currently, many computing solutions, especially for applications in the scientific area, are developed by taking advantage of thousands of cores available in the multiprocessors found in a board video, due to technologies such as NVIDIAs Compute Unified Device Architecture (CUDA) and Open Computing Language (OpenCL) [5].

The implementation of Evolutionary Particle Swarm Optimization Algorithm in a cooperative and multi-swarm model applied under GPGPU paradigm (general-purpose computing on graphics processing units) has a number of benefits, since each element can be treated by a thread, which tends to contribute in reduction of execution time, in addition to the potential increase in performance of search and optimization.

This paper presents an algorithm based on both EPSO and PSO mechanisms over a cooperative approach of master and slaves swarms implemented under CUDA architecture. This new algorithm is called Cooperative Evolutionary Multi-Swarm Optimization on Graphics Processing Units (CEMSO-GPU).

For comparison and validation, we used four engineering problems with large usage in scientific literature: Welded Beam Design (WBD); Air Tank Design (ATD); Minimization of the Weight of a Tension/Compression Spring (MWTCS); Speed Reducer Design with 11 Restrictions (SRD-11).

2 Related Works

Some publications involving the use of PSO on GPUs were found in the scientific literature. The most relevant works for the context of this paper are briefly discussed below:

1. *Collaborative Multi-Swarm PSO for Task Matching Using Graphics Processing Units* [6]: in this work, the authors expose an implementation of the cooperative and multi-swarm model for PSO which was executed over CUDA architecture for a task matching operation. This work uses a multi-swarm environment model, similar to [4]. The algorithm uses data of discrete and continuous type and the entire process is carried out continuously. In the end, the authors apply rounding to discrete values. The results obtained in this study show considerable improvements in processing time and in the values obtained at the end of execution when compared to a single swarm PSO.
2. *MCPSO—A Multi-population Cooperative Particle Swarm Optimization* [13]: this paper presents a multi-swarm and cooperative PSO, which is used as basis for CEMSO-GPU approach. The MCPSO algorithm is a masterslave model that consists of one master swarm and several slave swarms. The evolution of slave swarms is likely to amplify the diversity of individuals of the population and consequently to generate more promising particles for the master swarm. The master swarm updates the particle states based on both its own experience and that of the most successful particles in the slave swarms.
3. *GPU-based Asynchronous Particle Swarm Optimization* [7]: this work presents an approach for asynchronous execution of PSO on the GPU and exposes the problems of a synchronous implementation. In this implementation, the authors propose a data processing where each thread represents a specific variable from the particle, and it is treated in a loop. The obtained results, in relation to the sequential version show an increase of performance of approximately 300-fold in tests using functions such as Rastrigin. No tests with restrictive functions were presented in this paper.

We conducted an extensive search in literature regarding the publications involving the EPSO algorithm, however, no work with respect to implementation of the method created by [3] on GPUs under CUDA architecture and its use in an multi-swarm approach has been found.

3 A Brief Description of Classic and Evolutionary Particle Swarm Optimization For CEMSO

3.1 Particle Swarm Optimization

PSO is a technique for stochastic optimization of nonlinear and continuous functions developed by James Kennedy and Russell Eberhart [8]. It was based from studies related to the "social behavior" observed in some species of birds. The PSO algorithm simulates a population of particles called swarm, which operates inside of a search space predetermined [9, 10]. A particle is represented by two vectors storing values of the position and speed, where the value of the number of vector elements is equal to the amount of N variables corresponding to a N-dimensional search space. Based on a collective and cooperative exploration, the swarm tend to progress in the search process as new and best results are found.

The swarm's beahvior is defined by the Eqs. (1)–(3), where Eq. (1) is the velocity update, Eq. (2) is the position update and Eq. (3) is the inertia weight.

$$V_i^{(t+1)} = wV_i^{(t)} + R_1C_1(b_i - X_i^{(t)}) + R_2C_2(b_g - X_i^{(t)}) \tag{1}$$

$$X_i^{(t+1)} = X_i^{(t)} + V_i^{(t+1)} \tag{2}$$

$$w = \frac{(k-1)}{(I_{MAX}-1)(-W_{MAX}+W_{MIN})+W_{MAX}} \tag{3}$$

In order to avoid particles from escaping the search space, the position values of the particle are subjected to a modified damping boundary condition process. This feature is used only if the position value is outside the minimum or maximum limits. Another important improvement added to the boundary conditions is a new disturbance variable called α constant, which helps to prevent particles that may get trapped in local best by multiplying a uniform random number between 0 and 1 with the α constant, follow by an addition or subtraction with the maximum or minimum limit respectively. It is important to note that if the values of position of the particles exceed the limits, the speed is also changed to the negative value of its current value and multiplied by a random number generated between 0 and 1. The scheme for boundary conditions with α constant on the particles velocity and position are described by Eqs. (4) and (5).

$$V_i^{(t+1)} = \begin{cases} -V_i^{(t+1)}R_1 & \text{if } X_i^{(t+1)} < X_{MIN} \\ V_i^{(t+1)} & \text{if} X_{MIN} \le X_i^{(t+1)} \le X_{MAX} \\ -V_i^{(t+1)}R_1 & \text{if } X_i^{(t+1)} > X_{MAX} \end{cases} \tag{4}$$

$$X_i^{(t+1)} = \begin{cases} X_{MIN} + R_2\alpha & \text{if } X_i^{(t+1)} < X_{MIN} \\ X_i^{(t+1)} & \text{if } X_{MIN} \leq X_i^{(t+1)} \leq X_{MAX} \\ X_{MAX} - R_2\alpha & \text{if } X_i^{(t+1)} > X_{MAX} \end{cases} \quad (5)$$

The particles movement submitted to a customized damping boundary condition process with α constant is shown in Fig. 1.

Table 1 describes the variables found on Eqs. (1)–(5).

The PSO algorithm used as a basis for implementation of the CEMSO-GPU algorithm is described in Algorithm 1:

The PSO algorithm used as a basis for implementation of the CEMSO-GPU algorithm is described in Algorithm 1:

```
Initialize the population (velocity and positions);
Evaluate fitness for each particle in the swarm;
Initialize the local best (b_i);
Initialize the global best (b_g);
for k ← 1 to (I_MAX − 1) do
    Update inertia weight using equation (3);
    forall the Particles in the swarm do
        Update velocity using equation (1);
        Update position using equation (2);
        Apply velocity correction with boundary conditions using equation (4);
        Apply position correction with boundary conditions using equation (5);
        Update fitness for each particle;
        Update particles local best (b_i);
        Update swarms global best (b_g);
    end
end
```

Algorithm 1: Example of a classic PSO algorithm with inertia weight

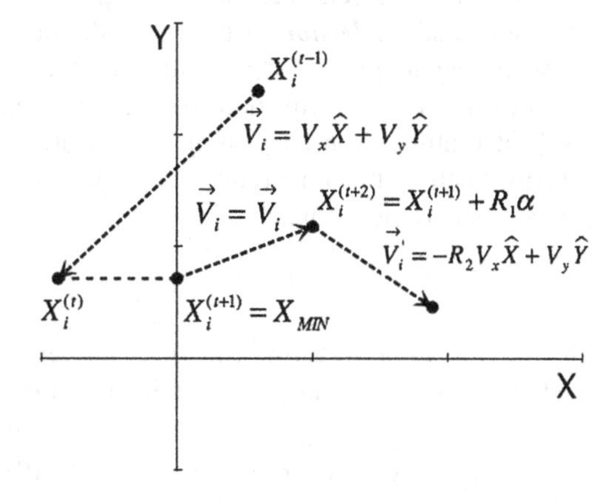

Fig. 1 Example of a custom damping scheme for a two-dimensional problem with α constant

Table 1 List of variables for PSO and its variants

Variable	Description
t	Time step
i	Particle's index
w	Inertia weight, with a linear decrease at each iteration
V	Particle's velocity
X	Particle's position
C_1	Acceleration constant for individual interaction
C_2	Acceleration constant for social interaction
b_i	Particle's local best
b_g	Swarm's global best
$R_1 \| R_2$	Uniform random numbers between 0 and 1
$W_{MAX} \| W_{MIN}$	Inertia weight's maximum and minimum values
I_{MAX}	Iteration's maximum value
k	Iteration index
$X_{MAX} \| X_{MIN}$	Maximum and minimum values for particle's position
a	Disturbance constant for boundary conditions

3.2 Evolutionary Particle Swarm Optimization

The EPSO algorithm, developed by Miranda and Fonseca [3], is a metaheuristic that adds the mechanism of evolutionary strategies (EE) to the PSO algorithm, where the operators of classical recombination are replaced by rules of motion of the particles. According to [3], EPSO can be classified in two forms, given the mechanisms of the algorithm, it can be interpreted as a variant of the PSO or as a variant of EE.

The mechanisms of evolutionary strategy found in EPSO of [3, 11] are: ***Replication*** of particles; ***mutation*** of the following weights: inertia weight, acceleration constants (C_1 and C_2) and the global best's position values; ***reproduction*** of new particles described by Eq. (6) with weights mutated; ***evaluation*** of new individuals; ***selection*** of the best individuals.

According to [3, 11] the addition of the evolutionary mechanisms to PSO provides a search system more robust, since the mutation can produce a better result at runtime. The changes in the equation of velocity proposed by [3] are described below. It is important to note that the equations of inertia weight (3) and position (2) are not changed.

$$V_i^{(t+1)} = mw^* V_i^{(t)} + mC_{1(i)}^*(b_i - X_i^{(t)}) + mC_{2(i)}^*(b_g^* - X_i^{(t)}) \tag{6}$$

The differences between the equations for velocity (1) and (6) are the absence of random number generators and the inclusion of the mutation process to the variables mw, $mC_{1(i)}^*$, $mC_{2(i)}^*$ and the position values from global best (b_g^*). The mutation process are expressed by Eqs. (7)–(10).

$$mw^{*}_{(i)} = w + (1 + \sigma N(0,1)) \tag{7}$$

$$mC^{*}_{1(i)} = C_1 + (1 + \sigma N(0,1)) \tag{8}$$

$$mC^{*}_{2(i)} = C_2 + (1 + \sigma N(0,1)) \tag{9}$$

$$b^{*}_{g} = b_g + (1 + \sigma_g N(0,1)) \tag{10}$$

The new variables found in Eqs. (6)–(10) are described in Table 2.

Given the description above, the customized EPSO algorithm (based on the work developed by [3]) used as the basis for implementation of CEMSO-GPU algorithm is described in the Algorithm 2:

```
Initialize the population (velocity and positions);
Evaluate fitness for each particle in the swarm;
Initialize the local best (b_i);
Initialize the global best (b_g);
for k ← 1 to (I_MAX − 1) do
    Update inertia weight using equation (3);
    foreach Particles in the swarm do
        Update velocity from the original particles using equation (1);
        Update position from the original particles using equation (2);
        Apply velocity correction with boundary conditions to the original
        particles using equation (4);
        Apply position correction with boundary conditions to the original
        particles using equation (5);
        Update fitness for the original particles;
        Update particles local best (b_i);
        Update swarms global best (b_g);
        Replicate particle N times;
        foreach Particle's Replicate do
            Apply mutation for each weights (w, C_1, C_2) to all replicated
            particles;
            Update velocity from the replicated particles using equation (6);
            Update position from the replicated particles using equation (2);
            Apply velocity correction with boundary conditions to the
            replicated particles using equation (4);
            Apply position correction with boundary conditions to the
            replicated particles using equation (5);
            Evaluate fitness for the replicated particles;
        end
        Select the best particles;
        Update particles local best (b_i);
        Update swarms global best (b_i);
    end
end
```

Algorithm 2: Example of a evolutionary PSO (EPSO) algorithm with inertia weight

Table 2 List of variables for EPSO

Variable	Description
mw	Inertia weight, submitted to the mutation process
$mC^*_{1(i)}$	C_1 constant submitted to the mutation process
$mC^*_{2(i)}$	C_2 constant submitted to the mutation process
b_g	Swarm's global best
$*$	Marker of the variables subjected to the mutation process
σ_g	Disturbance constant for the global best
σ	Mutation parameter for mw, $mC^*_{1(i)}$ and $mC^*_{2(i)}$
$N(0, 1)$	Gaussian distribution of mean 0 and standard deviation 1

4 Compute Unified Device Architecture (CUDA)

CUDA architecture provides a set of extensions to the C language, which allow the implementation of parallel algorithms in video cards. GPUs with CUDA architecture have many cores which allows to run collectively thousands of independent small parts of processing, called threads [5].

The CUDA SDK (Software Development Kit) includes a compiler adapted for heterogeneous computing paradigm GPGPU (General Purpose Computing on Graphics Processing Units) and other tools capable of supporting heterogeneous applications, which have serial and parallel parts performed on the CPU (host) and GPU (device), respectively [12].

4.1 Thread Organization

To keep the organization in the executing data parallelism and data distribution effectively with control of memory access, CUDA uses three levels of organization: thread, block and grid. The explanation of each one follows:

1. *Thread*: It is the basic unit of execution. Each thread is responsible for a copy of the program for an user-determined amount of data. In some solutions, it is possible each thread remains responsible for an address of a vector or a matrix;
2. *Block*: This structure stores a vector or a matrix (2D or 3D) of threads. All blocks have the same quantities of threads. Through this organization, it is possible to use synchronization of threads to the block level;
3. *Grid*: It is the set of all threads that a kernel will use. A Grid is a vector or matrix of blocks. In multi-GPU environments, the Grids can not exchange information among themselves.

Figure 2 shows the schematic organization of threads and blocks on CUDA architecture, where the parallel application (executed into a function called kernel) uses a quantity of four blocks with five threads each.

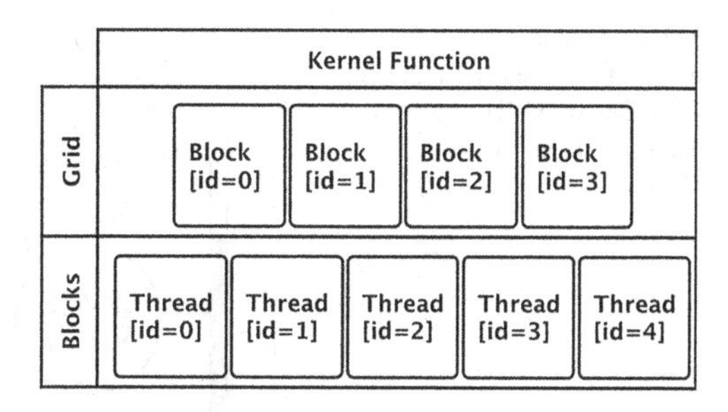

Fig. 2 Scheme of threads and blocks. Adapted from [5]

5 Cooperative Multi-Swarm Approach for EPSO on the CUDA Architectur (CEMSO-GPU)

The CEMSO algorithm is a result of the integration between the mechanisms found in EPSO and PSO, acting under a multi-swarm approach.

It consists of an environment where two or more slave swarms cooperate with the master swarm the values of the particles classified as being the global best found to current iteration. The optimization process that occurs in the master swarm tends to enhance the global best of the slave swarms, and contributes in finding the global best in relation to the master swarm.

The use of CUDA architecture for CEMSO algorithm allows the slave swarm to be executed in parallel, massively and with a low processing time. By comparison with other parallel and distributed computing methods, such as MPI (Message Parsing Interface), CUDA has the advantage of running many threads (SIMD taxonomy) on low cost computers with no need for clusters. In relation to the parallel scheme for slave swarms, the CUDA architecture provides a programming environment that makes possible to execute EPSO and PSO algorithms with high level of parallelism.

5.1 Multi-Swarm Structure with Slave/Master Approach

The basic structure of the multi-swarm environment in CEMSO algorithm is based on the model found in the MCPSO algorithm [13]. This approach is organized so that the slave swarms are executed in parallel, without exchanging information among themselves, where they provide their best solutions (global best) to the master swarm. The cooperativity of data is applied only in the relationship between the master swarm and the slave swarms.

The master swarm updates its particles, taking into consideration the information of the best solution found among all slave swarms. Note that this information is integrated as a third component in the velocity update equation (more details in Sect. 5.2). Figure 3 shows the architecture of CEMSO algorithm based on the model proposed by [13] and adapted for CUDA.

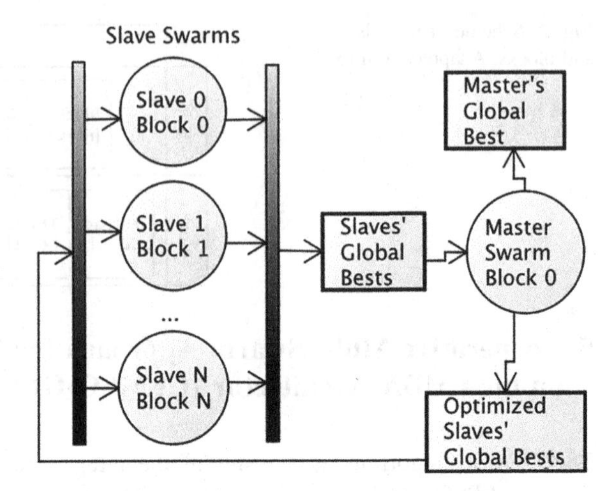

Fig. 3 Schematic illustration of the master/slave model for the CEMSO-GPU algorithm. Modified from [13]

Based on the parallel implementation in CUDA, each thread is responsible for the manipulation of a particle (one thread, one particle (cardinality 1:1)), each block being represented by a swarm (one block, one swarm (cardinality 1:1)).

Figure 4 shows the arrangement of particles per thread, slave ande master swarms per grid in the CEMSO-GPU algorithm.

5.2 Improvements on Equations

Slave Swarms The slave swarms in CEMSO-GPU algorithm operates in a parallel and an independent way in the search and optimization of the problem and they all share the same amount of particles. Basically, each swarm executes in parallel the process described in Algorithm 2 with some changes.

A few important points to be highlighted about the velocity update for CEMSO are described below:

1. The updating equations for velocity and position of original particles are not changed (Eqs. (1) and (2) respectively);
2. The velocity update equation applied to the replicas is also modified in order to add a uniform random number generator in a range between zero and one (R_1 and R_2). The addition of these variables allows a higher exploration in the search space by the replicas.

Changes in the Eq. (6) for replicas in the slave swarms can be seen in Eq. (11).

$$V_i^{(t+1)} = mw^* V_i^{(t)} + R_1 mC^*{}_{1(i)}(b_i - X_i^{(t)}) + R_2 mC^*_{2(i)}(b_g^* - X_i^{(t)}) \quad (11)$$

Master Swarm The use of the master swarm under the context of the CEMSO algorithm occurs after the parallel execution of slave swarms, where the values of the global best from every slave (b_g^S), as well as, the values of velocity and fitness of each slave swarm are copied.

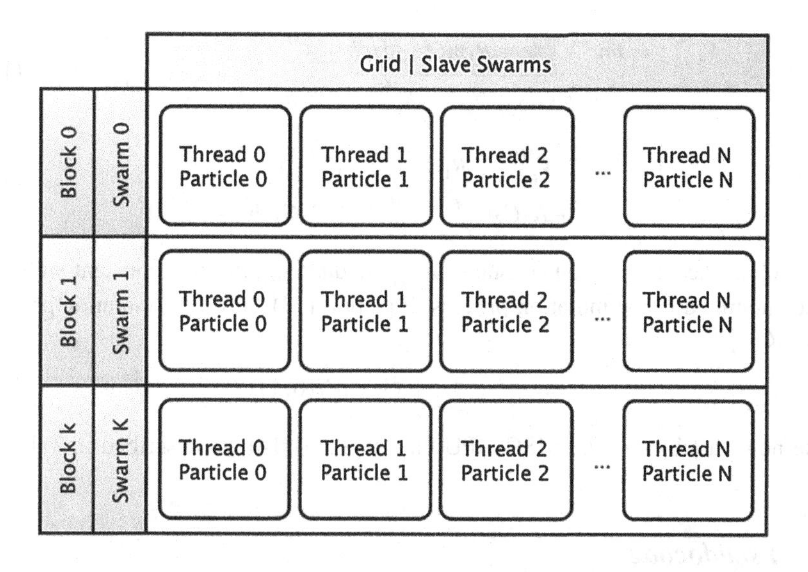

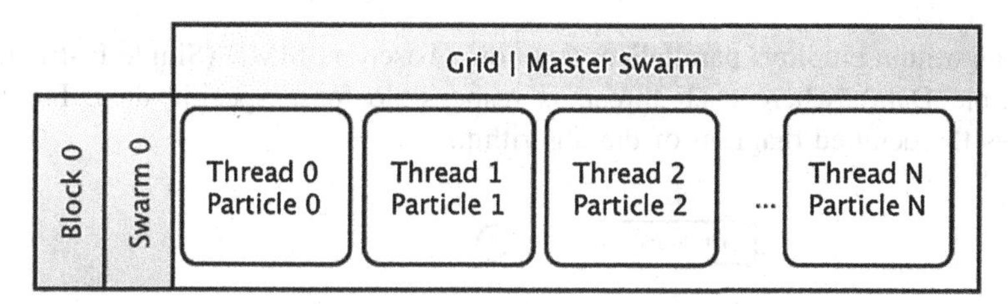

Fig. 4 Scheme of the slave and master swarms in CUDA

The update equations for velocity proposed in [13] are used in the master swarm, where the acceleration constant C_3 and the values of the best minimal/ maximum global found by the slave swarms (b_g^S) are added. It is important to emphasize that the equation for velocity used on replicated particles, Eq. (11), was also adapted for the equation proposed by [13], where the mutated factors were added. Changes in the velocity equations are described in the Eqs. (12) and (13).

Table 3 List of variables for CEMSO-GPU

Variable	Description
C_3	Acceleration constant for master swarm's social interaction
$mC_{3(i)}$	C3 constant submitted to the mutation process
$R1\|R2\|R3$	Uniform random numbers between 0 and 1
b_i^M	Master swarms local best
b_g^M	Master swarm's global best
b_g^S	Best global value obtained by the slave swarms

$$V_i^{(t+1)} = mw^* V_i^{M(t)} + R_1 mC^*{}_{1(i)}(b_i^M - X_i^{M(t)}) + R_2 mC^*{}_{2(i)}(b_g^{M*} - X_i^{M(t)}) + R_3 mC^*{}_{3(i)}(b_g^S - X_i^{M(t)}) \quad (12)$$

$$V_i^{(t+1)} = wV_i^{M(t)} + R_1 C_1(b_i^M - X_i^{M(t)}) + R_2 C_2(b_g^M - X_i^{M(t)}) + R_3 C_3(b_g^S - X_i^{M(t)}) \quad (13)$$

As well as the acceleration constants $mC_{1(i)}$ and $mC_{2(i)}$, the new constant $mC_{3(i)}$ is also submitted to the mutation process. Equation (14) shows the mutation process for $mC_{3(i)}$.

$$mC^*_{3(i)} = C_3 + (1 + \sigma N(0, 1)) \quad (14)$$

The new variables for CEMSO-GPU on Eqs. (12)–(14) are described in Table 3.

5.3 Pseudocode

The algorithm employs parallelism taxonomy based on SIMD (Single Instruction, Multiple Data), where each thread is responsible for a specific data. Figure 5 shows the detailed diagram of the algorithm.

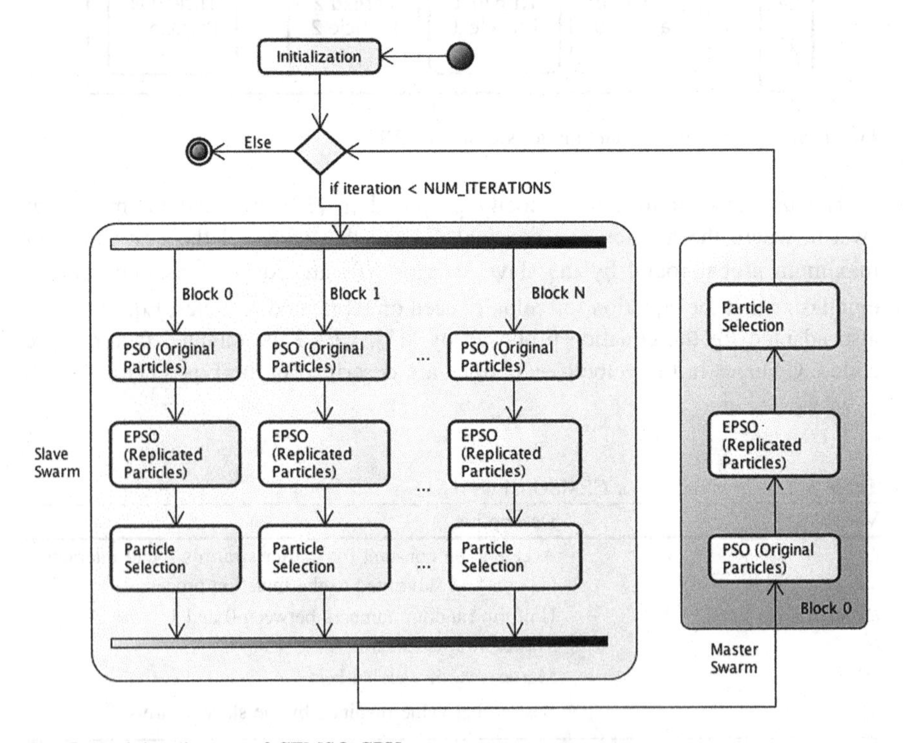

Fig. 5 Activity diagram of CEMSO-GPU

The CEMSO pseudocode is divided in three algorithms: Algorithm 3 describes the initalization, Algorithm 4 shows the execution of slave swarms and Algorithm 5 exposes the master swarm execution flow.

```
Allocate memory space for reading/writing on the GPU;
Run in Parallel For All Slave Swarms (multi-Block)
|   Initialize the population (velocity and positions);
|   Evaluate fitness for each particle in the swarm;
|   Initialize the local best (b_i);
|   Initialize the global best (b_g);
|   Sync threads (wait for all threads to finish);
|   if threadIndex=0 then
|   |   Initialize global best (b_g) of each slave swarm;
|   end
end
Sync blocks (wait for all blocks to finish);
```

Algorithm 3: CEMSO-GPU - Initialization

```
for k ← 1 to (I_MAX − 1) do
|   Update inertia weight using equation (3);
|   Run in Parallel For All Slave Swarms (multi-Block)
|   |   Update velocity from the original particles using equation (1);
|   |   Update position from the original particles using equation (2);
|   |   Apply velocity correction to the original particles using equation (4);
|   |   Apply position correction to the original particles using equation (5);
|   |   Update fitness for the original particles;
|   |   Update particles local best (b_i);
|   |   Sync threads (wait for all threads to finish);
|   |   if threadIndex=0 then
|   |   |   Update global best (b_g) of each slave swarm;
|   |   end
|   |   Sync threads (wait for all threads to finish);
|   |   Replicate particle N times;
|   |   foreach Particle's Replicate do
|   |   |   Apply mutation for each weights (w, C_1, C_2) from all replicated
|   |   |   particles;
|   |   |   Update velocity from the replicated particles using equation (11);
|   |   |   Update position from the replicated particles using equation (2);
|   |   |   Apply velocity correction to the replicated particles using equation
|   |   |   (4);
|   |   |   Apply position correction to the replicated particles using equation
|   |   |   (5);
|   |   |   Update fitness for the replicated particles;
|   |   |   Select the best particles for the next iteration (slave swarm);
|   |   end
```

Update particles local best (b_i);
Sync threads (wait for all threads to finish);
if *threadIndex=0* **then**
Update global best (b_g) of each slave swarm;
end
end
end

Algorithm 4: CEMSO-GPU - Slave Swarms

for $k \leftarrow 1$ **to** $(I_{MAX} - 1)$ **do**
Sync blocks (wait for all blocks to finish);
Update the best global value found in the slave swarms (b_g^S);
Send to the master swarm every slave swarms global best (b_g);
Run in Parallel For Master Swarm (single-*Block*)
Update velocity from the original particles using equation (13);
Update position from the original particles using equation (2);
Apply velocity correction to the original particles using equation (4);
Apply position correction to the original particles using equation (5);
Update fitness for the original particles;
Update particles local best (b_i^M);
Sync threads (wait for all threads to finish);
if *threadIndex=0* **then**
Update masters global best (b_g^M);
end
Sync threads (wait for all threads to finish);
Replicate particle N times;
foreach *Particle's Replicate* **do**
Apply mutation for each weight (w, C1, C2, C3) from all replicated particles;
Update velocity from the replicated particles using equation (12);
Update position from the replicated particles using equation (2);
Apply velocity correction to the replicated particles using equation (4);
Apply position correction to the replicated particles using equation (5);
Update fitness for the replicated particles;
Select the best particles for the next iteration (master swarm);
end
Update particles local best (b_i^M);
Sync threads (wait for all threads to finish);
if *threadIndex=0* **then**
Update masters global best (b_g^M);
end
end
Sync blocks (wait for all blocks to finish);
end

Algorithm 5: CEMSO-GPU - Master Swarm

6 Experiments

For comparison, the CEMSO algorithm was subjected to tests with four engineering problems widely used in scientific literature: Welded Beam Design (WBD); Speed Reducer Design with 11 restrictions (SRD-11); Air Tank Design (ATD); Minimization of the Weight of a Tension/Compression Spring (MWTCS). The algorithm CEMSO-GPU is written in CUDA-C programming language, which allows the use of the GPU for general purposes.

The parameter values are: Maximum number of iterations = 1000; number of particles for each slave swarm = 40; number of particles for master swarm = 10; number of replicates per particle = 5; number of slave swarms = 10; disturbance constant for boundary conditions $(\alpha) = 0.22$; disturbance constant to the global best $(\sigma_g) = 0.005$; parameter of strategies for mutation of the inertia and acceleration factors $(\sigma) = 0.22$; factors C_1 and $C_2 = 2.05$; acceleration factor for master swarm $(C_3) = 2.02$; number of executed tests per function = 20. The results shown in Tables 4, 5, 6 and 7 are the best values found in 20 executions for each problem.

The hardware description is shown on the list below:

1. *CPU*: Intel Core i5 with 2.66 GHz and 4 GB of memory RAM DDR3;
2. *GPU*: NVIDIA GeForce GT 330M with 256 MB VRAM GDDR3, 48 CUDA cores and compute capability of 1.2;

6.1 Engineering Optimization Problems

Welded Beam Design (WBD) Minimizing the manufacturing cost of a steel beam, subject to some restrictions, such as: Shear stress, bending stresses on the beam, buckling on the bar and deflection of the beam end and side constraints [15]. The parameters for this problem are: Thickness of the solder (*H*); beam lenght (*L*); width of the beam (*T*), length of the weld (*B*). Figure 6 shows the layout of WBD problem.

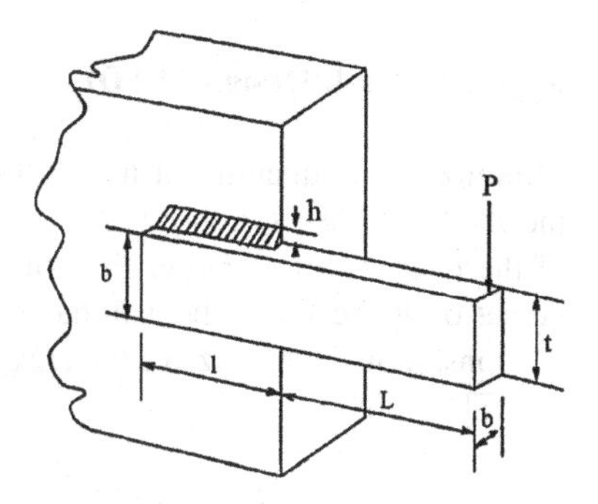

Fig. 6 Scheme of Welded Beam Design (WBD). *Source* [15]

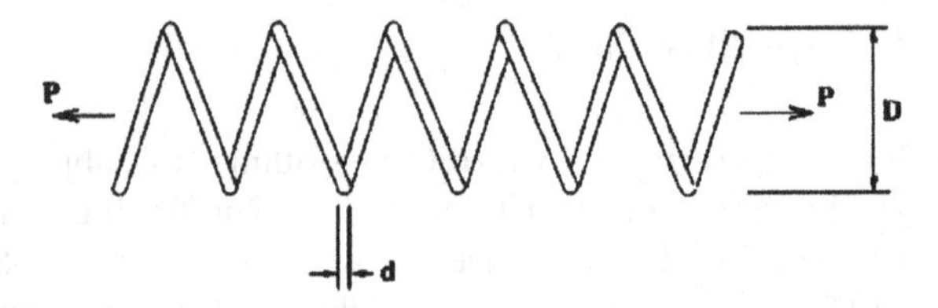

Fig. 7 Scheme of minimization of the Weight of a Tension/Compression Spring (MWTCS). *Source* [18]

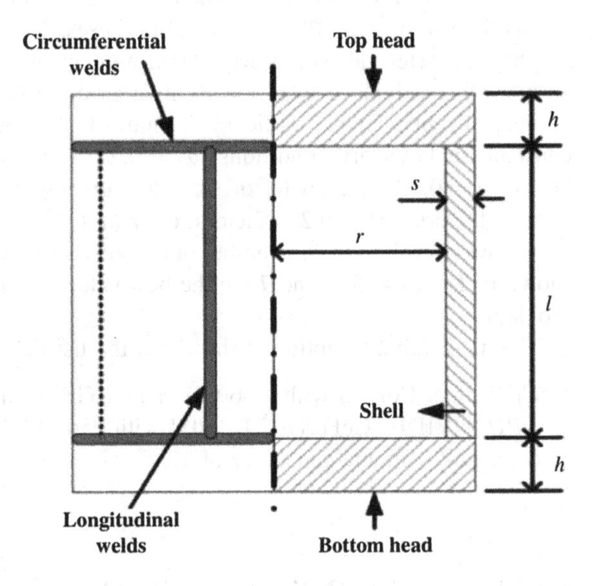

Fig. 8 Scheme of Air Tank Design (ATD). *Source* [18]

Minimization of the Weight of a Tension/Compression Spring (MWTCS) Minimizing the weight of the tension/compression of a spring, subject to some restrictions, such as: minimum deflection, shear stress, the wave frequency, limits on outside diameter and design variables [15]. The parameters for this problem are: wire diameter (d); diameter of spring (D); number of active coils (P). The constraints for MWTCS problem are represented by G variables (A total of four constraints). Figure 7 shows the layout of MWTCS problem.

6.1.1 Air Tank Design (ATD)

Minimizing the quantity of material used, which depends on the inner radius (r), the shell thickness (s), the shell length (l), and the head thickness (h). The volume of the tank has to be larger than the specified volume (constraint $G1$), the thicknesses of the head and the wall have to satisfy the ASME code ($G2$, $G3$), and there are constraints on the size of the tank ($G4$, $G5$, $G6$) [18]. Figure 8 shows the layout of ATD problem.

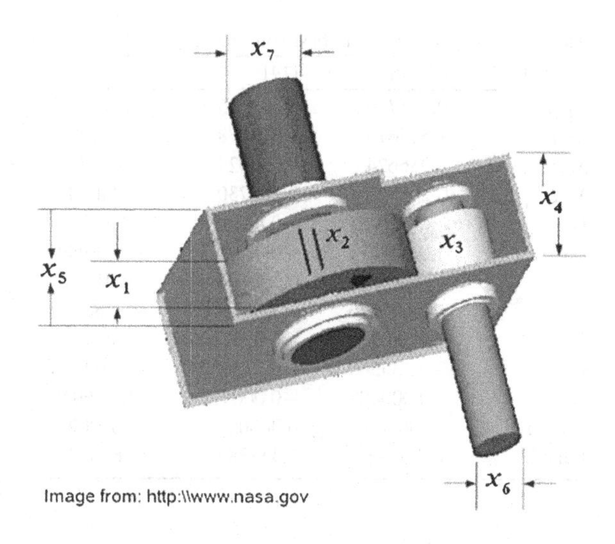

Fig. 9 Scheme of Air Tank Design (ATD). *Source* [18]

6.1.2 Speed Reducer Design for 11 Restrictions (SRD-11)

Speed reduction system by minimizing the weight subject to certain restrictions, such as: Bending stress of gear teeth, surface tension, transverse deviations of the stems and tensions on the axis [15]. The variables of the problem are: width face

Table 4 Comparison of the best solutions for SRD-11

Variables	CEMSO	[15]	[20]	[21]
X_1	3.500000	3.500459	3.500000	3.500000
X_2	0.700000	0.700020	0.700000	0.700000
X_3	17.000000	17.00503	17.000000	17.000000
X_4	7.300000	7.300251	7.300000	7.300000
X_5	7.800000	7.800195	7.800000	7.800000
X_6	2.900000	2.900041	3.350215	3.350214
X_7	5.286684	5.286863	5.286683	5.286683
G_1	−0.073915	−0.074364	−0.073915	−0.073915
G_2	−0.197999	−0.198624	−0.197996	−0.197998
G_3	−0.107955	−0.108202	−0.499172	−0.499172
G_4	−0.901472	−0.901443	−0.901471	−0.901471
G_5	−1.000000	−1.000000	−2.22E-16	0.000000
G_6	0.000000	−0.000102	−3.331E-16	−5.00E-13
G_7	−0.702500	−0.702403	−0.702500	−0.702500
G_8	0.000000	−0.000103	0.000000	−1.00E-16
G_9	−0.795833	−0.795801	−0.583333	−0.583333
G_{10}	−0.143836	−0.143857	−0.051326	−0.051325
G_{11}	−0.010852	−0.011074	−0.010852	−0.010852
Violations	0	0	N/A	N/A
Fitness	2896.259277	2897.531422	2996.348165	2996.348165

Table 5 Comparison of the best solutions for WBD

Variables	CEMSO	[14]	[15]	[16]	[17]
$X_1(h)$	0.205730	0.205735	0.171937	0.202369	0.205730
$X_2(l)$	1.517675	1.517678	4.122129	3.544214	3.470489
$X_3(t)$	9.036624	9.036624	9.587429	9.048210	9.036624
$X_4(b)$	0.205730	0.205730	0.183010	0.205723	0.205729
G_1	−0.001953	−0.001953	−8.067400	−12.839796	0.000000
G_2	−0.001953	−0.001953	−39.336800	−1.247467	0.000002
G_3	0.000000	0.000000	−0.011070	−0.001498	0.000000
G_4	−3.607646	−3.607646	−3.467150	−3.429347	−3.432984
G_5	−0.080730	−0.080730	−0.236390	−0.079381	−0.080730
G_6	−0.235540	−0.235540	−16.024300	−0.235536	−0.235540
G_7	−0.000488	−0.000977	−0.046940	−11.681355	0.000001
Violations	0.000000	0.000000	0.000000	N/A	2
Fitness	1.458883	1.458885	1.664373	1.728024	1.724852

Table 6 Comparison of the best solutions for ATD

Variables	CEMSO	[18]
$X_1(h)$	6.490450	13.67
$X_2(l)$	603.048584	610.00
$X_3(r)$	49.898628	105.18
$X_4(s)$	0.600016	1.01
G_1	−0.000035	−0.000000001
G_2	−0.000559	0.0000917*
G_3	−0.202475	0.00000000022*
G_4	−0.983418	−0.98
G_5	−0.663342	−0.29
G_6	−0.011396	0.000000000783*
Violations	0.000000	0.000000
fitness	218122.468750	1380000 (1.38×10^6)

Table 7 Comparison of the best solutions for MWTCS

Variables	CEMSO	[15]	[16]	[22]
$X_1(d)$	0.050000	0.050180	0.051728	0.051989
$X_2(D)$	0.282023	0.279604	0.357644	0.363965
$X_3(P)$	2.000000	2.087959	11.244543	10.890522
G_1	0.000000	−0.002840	−0.000845	−0.000013
G_2	−0.235327	−0.249450	−0.0000126	−0.000021
G_3	−43.146137	−42.176000	−4.051300	−4.061338
G_4	−0.778651	−0.780140	−0.727090	−0.722698
Violations	0.000000	0.000000	N/A	N/A
Fitness	0.002820	0.002878	0.0126747	0.012681

($X1$); teeth module($X2$); number of teeth of the pinion ($X3$); length of the first axis between the bearings ($X4$); length of the second axis between the bearings ($X5$); diameter of the first axis ($X6$); diameter of the second axis ($X7$). Figure 9 shows the layout of the SRD-11 problem.

6.2 Results

In [18], constraints are satisfied with a tolerance of 0.01 % of the initial values of the constraints.

7 Conclusion and Future Works

Based on the results obtained from the experiments involving the engineering problems, the CEMSO-GPU algorithm showed the best results compared to the values obtained by other implementations. Although the focus of this work has been the implementation of the algorithm under CUDA, CEMSO-GPU can be easily ported to other parallel architectures such as Beowulf with MPI. Some key points about the experiments can be seen below:

1. The improvement noted in the results can be verified and attributed by following factors: (1) The use of boundary conditions with correction factor applied to a random variable with the rate of correction, where the particle that exceeds the limits of search returns to inner search space; (2) The execution of the master swarm in the particles tends to produce optimized values for each best global of the slave swarms; (3) The replicates produced by each particle allowed for a better exploration of the search space;
2. From a simplified viewpoint, the algorithm CEMSO can be classified as a hybrid metaheuristic between PSO and EPSO applied under a multi-swarm approach.
3. The use of CUDA had a significant contribution to improvements on performance and feasibility of execution for multi-populations algorithms with a small time processing (average of 2.1472 s for WBD, 2.0155 s for ATD, 1.9833 s for MWTCS and 2.4109 s for SRD-11 among the twenty executions of each problem). In a final analysis, the GPGPU platform of massive parallelism opens the possibility for testing which involves large loads of processing data in a feasible runtime.

As some future works, we can highlight some key points:

1. Inclusion of mechanisms for Social Interaction based on Game Theory, inspired on the work developed by [15];
2. More research related to other methods involving boundary correction;

3. Study and implementation of mechanisms from other metaheuristics to CEMSO (e.g. Genetic Algorithms (GA), Artificial Immunological System Optimization (AISO), Ant Colony Optimization (ACO));
4. A technical comparison analysis involving other parallel methods and architectures (e.g. MPI, OpenMP, pThreads, etc.).

Acknowledgments This work is supported financially by Research Support Foundation of Par(FAPESPA) and Federal University of Pará (UFPA).

References

1. Bastos Filho, C.J.A., Caraciolo, M.P., Miranda, P.B.C., Carvalho, D.F: Multi ring PSO. In: The 10th Brazilian Symposium on Neural Networks (SBRN'2008), pp. 111–116 (2008)
2. Lopes, H.S., Takahashi, R.H.C.: Computação Evolucionária em Problemas de Engenharia, 1st edn. Ed. OMNIPAX, (2011) (in portuguese)
3. Miranda, V., Fonseca, N.: EPSO—Evolutionary particle swarm optimization, a new algorithm with applications in power systems. In: Transmission and Distribution Conference and Exhibition 2002: Asia Pacific. IEEE/PES, vol. 2, pp. 745–750 (2002)
4. Van Den Bergh, H., Engelbrecht, A.P.: A Cooperative approach to particle swarm optimization. IEEE Trans. Evol. Comput. **8**, 225–239 (2004)
5. Kirk, D.B., Hwu, W.M.: Programming Massively Parallel Processors a Hands-on Approach, 1st edn. Elsevier, Oxford (2010)
6. Solomon, S., Thulasiraman, P., Thulasiraman, R.: Collaborative multi-swarm PSO for task matching using graphics processing units. In: GECCO '11 Proceedings of the 13th Annual Conference on Genetic and Evolutionary Computation, vol. 2, pp. 1563–1570 (2011)
7. Mussi, L., Nashed, Y.S.G., Cagnoni, S.: GPU-based asynchronous particle swarm optimization. In: GECCO '11 Proceedings of the 13th Annual Conference on Genetic and Evolutionary Computation, vol. 2, pp. 1555–1562 (2011)
8. Kennedy, J., Eberhart, R.: Particle swarm optimization. In: Proceedings of the IEEE International Conference on Neural Networks, pp. 1942–1948 (1995)
9. Eberhart, R., Shi, Y.: Comparing inertia weights and constriction factors. In: Proceedings of the Congress on Evolutionary Computing, pp. 84–89 (2000)
10. Eberhart, R., Shi, Y.: A modified particle swarm optimizer. In: IEEE International Conference of Evolutionary Computation, pp. 69–73. Anchorage, Alaska (1998)
11. Leite, H., Barros, J., Miranda, V.: The evolutionary algorithm EPSO to coordinate directional overcurrent relay. In: 10th IET International Conference Developments in Power System Protection (DPSP 2010) Managing the Change, pp. 1–5 (2010)
12. Sanders, J., Kandrot, E.: CUDA by Example: An Introduction to General-Purpose GPU Programming. Addison-Wesley Professional, New York (2010)
13. Niu, B., Zhu, Y., He, X.: Multi-population cooperative particle swarm optimization. In: Proceedings of the European Conference on Artificial Life, pp. 874–883 (2005)
14. Souza, D.L., Monteiro, G.D., Martins, T.C., Teixeira, O.N., Dmitriev, V.A.: PSO-GPU: accelerating particle swarm optimization. In: CUDA-Based Graphics Processing Units. GECCO 2011, ACM Digital Library, pp. 837–838 (2011)
15. Teixeira, O.N., Lobato, W.A.L.L., Yanaguibashi, H.S., Cavalcante, R.V., Silva, D.J.A., Oliveira, R.C.L.: Algoritmo Genético com Interação Social na Resolução de Problemas de Otimização Global com Restrições (in portuguese), Computação Evolucionária em Problemas de Engenharia, Ed. OMNIPAX, 1st edn. pp. 197–223, (2011) (in portuguese)

16. He, Q., Wang, L.: An effective co-evolutionary particle swarm optimization for constrained engineering design problems. Eng. Appl. Artif. Intell. **20**, 89–99 (2007)
17. Mezura-Montes, E.: Coello Coello, C.: Useful infeasible solutions in engineering optimization with evolutionary algorithms. In: Proceedings of the 4th Mexican International Conference on Artificial Intelligence, MICAI 2005, Lecture Notes on Artificial Intelligence No. 3789, pp. 652–662 (2005)
18. Hsu, Y.L., Liu, T.C.: Developing a fuzzy proportional-derivative controller optimization engine for engineering design optimization problems. Eng. Optim. **39**(6), 679–700 (2007)
19. Golinski, J.: An adaptive optimization system applied to machine synthesis. Mech. Mach. Synth. **8**(4), 419–436 (1973)
20. Brajevic, I., Tuba, M., Subotic, M.: Improved artificial bee colony algorithm for constrained problems. In: Proceedings of the 11th WSEAS International Conference on Neural Networks, Fuzzy Systems and Evolutionary Computing, Stevens Point, USA: WSEAS, pp. 185–190 (2010)
21. Cagnina, L., Esquivel, S., Coello, C.: Solving engineering optimization problems with the simple constrained particle swarm optimizer. Informatica **32**(3), 319–326 (2008)
22. Coello C., Montes, E.: Constraint-handling in genetic algorithms through the use of dominance-based tournament selection. Adv. Eng. Inform. **16**, 193–203 (2002)

Hybrid Approach of Genetic Programming and Quantum-Behaved Particle Swarm Optimization for Modeling and Optimization of Fermentation Processes

Jun Sun, Vasile Palade, Zhenyu Wang and Xiaojun Wu

Abstract This paper proposes a novel method for modeling and optimization of fermentation process with a hybrid approach of genetic programming (GP) and quantum-behaved particle swarm optimization (QPSO). In this method, the GP algorithm is first used to model the process, with the parameters of the model selected randomly within a given interval, while the population of models evolves. Then, the parameters of the model obtained by the GP are tuned by the QPSO algorithm in order to increase the fitting accuracy. Finally, the values of the independent variables of the model representing the culture conditions are optimized by the QPSO in order to maximize the dependent variable, which generally represents the yield of the fermentation product. The proposed method is applied to the fermentation process of the hyaluronic acid (HA) production by *Streptococcus zooepidemicus*. The experimental results show the efficiency of the GP-QPSO approach in the modeling and optimization of this fermentation process.

Keywords Fermentation process · Genetic programming · Modeling · Optimization · Quantum-behaved particle swarm optimization

---

J. Sun (✉) · X. Wu
Key Laboratory of Advanced Control for Light Industry (Ministry of China), Jiangnan University, Wuxi, 214122 Jiangsu, China
e-mail: sunjun_wx@hotmail.com

X. Wu
e-mail: wu_xiaojun@yahoo.com.cn

V. Palade · Z. Wang
Department of Computer Science, University of Oxford, Wolfson Building, Parks Road, Oxford OX1 3QD, UK
e-mail: vasile.palade@cs.ox.ac.uk

Z. Wang
e-mail: zhen.yu.wang@cs.ox.ac.uk

I. Hatzilygeroudis and V. Palade (eds.), *Combinations of Intelligent Methods and Applications*, Smart Innovation, Systems and Technologies 23,
DOI: 10.1007/978-3-642-36651-2_7,

1 Introduction

The goal of a fermentation process is to produce various substances in the pharmaceutical, chemical and food industries. Its performance depends on many factors, including pH, temperature, ionic strengthens, agitation speed, and aeration rate in the aerobic fermentation [1]. To achieve a best performance of a fermentation process, various modeling and optimization strategies have been developed, the most frequently used one being "one-at-a-time" strategy [2]. This approach, however, is not only time consuming, but also ignores the combined interactions between physiochemical parameters [3]. In contrast, the response surface methodology (RSM), which includes factorial design and regression analysis, seeks to identify and optimize significant factors to maximize the response (cell density, high yields of the desired metabolic products or enzyme levels in the microbial system). RSM produces a model, which mathematically describes the relationship that exists between the independent and dependent variables of the process. The most widely used functions in the model developing stage of the RSM are second-order polynomials [4, 5]. The RSM has been widely applied in the modeling and optimization of biochemical processes [6, 7].

In this paper, a method based on genetic programming (GP) and quantum-behaved particle swarm optimization (QPSO) is proposed for modeling and optimization of fermentation processes. Crammer presented the first modern tree-based GP [8] in 1985 and the method was later extended and popularized by Koza [9]. Various applications of GP for several complex optimization and search problems [10–13] were examined. GP is well known to be computationally intensive and was originally used to solve relatively simple problems. The method has many novel and outstanding results recently in areas such as quantum computing, electronic design, game playing and sorting, due to improvements in GP technology and the exponential growth in CPU power [14–17]. The theory of GP has also received a formidable and rapid development in early 2000s. In particular exact probabilistic models, including schema theories and Markov chain models, have been built for GP.

The QPSO algorithm, a variant of particle swarm optimization (PSO), was inspired by quantum mechanics and the trajectory analysis of PSO [18, 19]. It uses a strategy based on a quantum δ potential well to sample around the previous best points [18, 19]. The QPSO algorithm essentially falls into the family of bare-bones PSO [20, 21], but uses double exponential distribution and an adaptive strategy to sample particle's positions. The iterative equation of QPSO is very different from that of PSO, and may lead QPSO to have improved search ability. Besides, unlike PSO, QPSO needs no velocity vectors for particles, and also has fewer parameters to adjust, making it easier to implement. The QPSO algorithm has been shown to successfully solve a wide range of continuous optimization problems and many efficient strategies have been proposed to improve the algorithm [22].

In the proposed method for modeling and optimization of fermentation processes, the GP is firstly used to model the fermentation process, with the parameters of the model (represented by an individual) being selected within a given interval during the evolution process of the GP. Then, the QPSO algorithm is employed to estimate the parameters of the model in order to improve fitting performance. After that, the QPSO is employed on the obtained model to optimize the culture conditions of the fermentation process, such that the fermentation production is maximized. The proposed GP-QPSO method is applied to model and optimize the hyaluronic acid (HA) production by *Streptococcus zooepidemicus*.

The rest of the paper is organized as follows. Section 2 provides introductions to the GP method and the QPSO algorithm. The proposed GP-QPSO approach is presented in Sect. 3. Section 4 gives the experimental results for the application of the GP-QPSO method to the HA production. Finally, the paper is concluded in Sect. 5.

2 Genetic Programming and Quantum-Behaved Particle Swarm Optimization

2.1 Genetic Programming (GP)

As a type of evolutionary algorithms (EAs), where each individual is a computer program, the GP is a methodology inspired by biological evolution to find computer programs that perform a user-defined task [23–25]. It is essentially a machine learning technique used to optimize a population of computer programs according to a fitness landscape determined by a program's ability to perform a given computational task.

The GP evolves computer programs, which are traditionally represented in memory as tree structures. Every tree node has an operator function and every terminal node has an operand, making mathematical expressions easy to evolve and evaluate. For example, Fig. 1 provides the tree structure of the mathematical expression:

$$\left(3 - \left(\frac{X}{11}\right)\right) + (7^{*} \cos(Y))$$

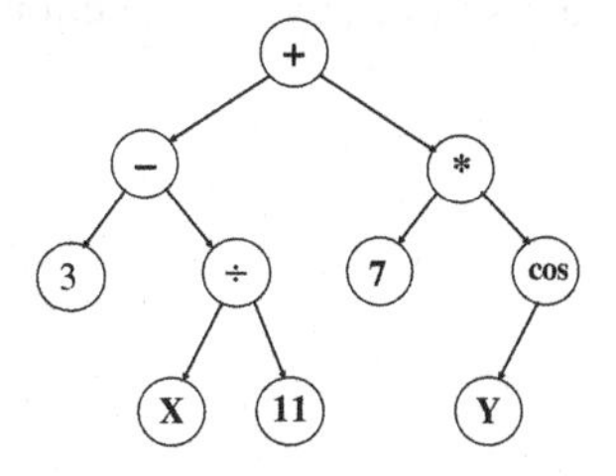

Fig. 1 An example of the tree structure of a mathematical expression

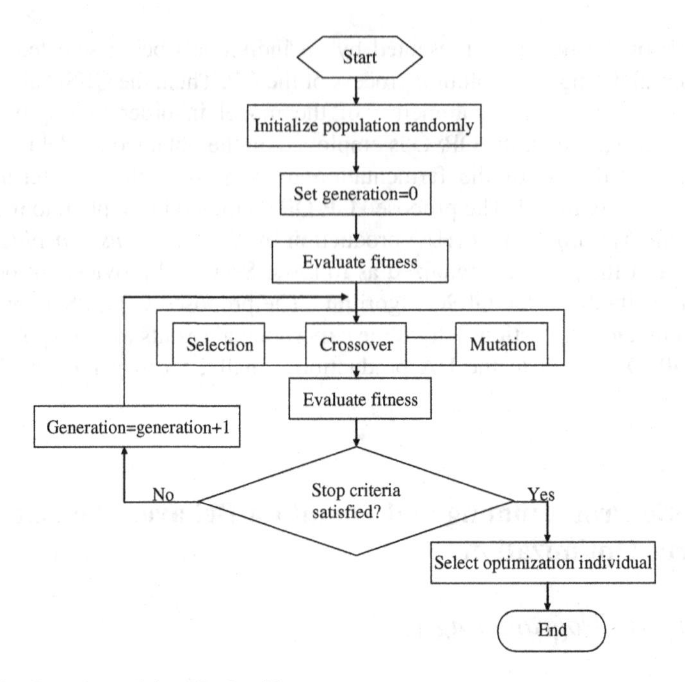

Fig. 2 The flow chart of the GP algorithm

The evolution in the GP, which is shown in the flow chart in Fig. 2, proceeds in similar way to a standard GA, i.e. an initial population is generated at random and each individual is evaluated to find its fitness value, and then it is evolved by means of genetic operators as follows.

2.1.1 Selection

Pairs of parent trees are selected based on its fitness for reproduction. Reproduction is where a selected individual copies itself into the new population. For the GP, the widely used methods of selection include fitness proportionate selection, tournament selection and rank selection. The most popular is fitness proportionate selection, which is employed in this paper. Under fitness proportionate selection, if $f(s_i(t))$ is the fitness of individual s_i in the population at generation t, the probability that individual s_i will be copied into the population of the next generation as a result of any reproduction is

$$p_{s_i}(t) = \frac{f(s_i(t))}{\sum_{i=1}^{M} f(s_i(t))}. \tag{1}$$

where M is the population size. When the reproduction operation is performed by means of the fitness-proportionate selection method, it is called fitness-proportionate reproduction.

2.1.2 Crossover

The crossover operation for the GP creates variation in the population by producing new offspring that consist of parts taken from each parent. The operation begins by independently selecting, using a uniform probability distribution, one random point in each parent to be the crossover point for that parent. The crossover fragment for a particular parent is the rooted subtree, whose root acts as the crossover point for that parent, and which consists of the entire subtree lying below the crossover point. The offspring is produced by swapping the crossover fragments of the parents at the crossover points.

2.1.3 Mutation

The mutation operation introduces random changes in the population. It begins by selecting a point at random for an individual tree structure. This mutation point can be an internal node (i.e., a function mode) or an external node (i.e., a terminal node) of the tree. The mutation operation then either replaces the selected node with its associated sub-trees generated randomly or changes its type.

2.2 The Quantum-Behaved Particle Swarm Optimization (QPSO)

2.2.1 A Brief Introduction to the PSO

The PSO algorithm was proposed originally by Kennedy and Eberhart as a population-based optimization technique [26], motivated by the social behavior of bird flocks or fish schooling. In PSO, the potential solutions, called particles, fly through the problem space by following their own experiences and the current best particle. Many empirical studies showed that the PSO algorithm is comparable in performance with, and can be considered as an alternative to the well known GA approach [27].

Since the origin of the PSO in 1995, the algorithm has gained increasing popularity during the last decade due to its effectiveness in performing difficult optimization tasks at a cheap computational cost. There has been a large amount of work done on the PSO algorithm, which involves theoretical analyses, improvements, and applications of the algorithm [28–39].

In the PSO with m individuals, each individual is treated as a volume-less particle in the D-dimensional space, with the current position vector and velocity vector of particle i at the kth iteration represented as $X_{i,k} = (X_{i,k}^1, X_{i,k}^2, \ldots, X_{i,k}^D)$ and $V_{i,k} = (V_{i,k}^1, V_{i,k}^2, \ldots, V_{i,k}^D)$. The particle moves according to the following equations:

$$V_{i,k+1}^j = w \cdot V_{i,k}^j + c_1 r_{i,k}^j (X_{i,k}^j - P_{i,k}^j) + c_2 R_{i,k}^j (X_{i,k}^j - G_k^j), \tag{2}$$

$$X_{i,k+1}^j = X_{i,k}^j + V_{i,k+1}^j, \tag{3}$$

For $i = 1, 2, \ldots m$; $j = 1, 2 \ldots, D$, where c_1 and c_2 are called acceleration coefficients. Parameter w is the inertia weight that can be adjusted to balance the exploration and exploitation of PSO [28]. Vector $P_{i,k} = (P_{i,k}^1, P_{i,k}^2, \ldots, P_{i,k}^D)$ is the best previous position (the position giving the best objective function value or fitness value) of particle i called *personal best* (*pbest*) position, and vector $G_k = (G_k^1, G_k^2, \ldots, G_k^D)$ is the position of the best particle among all the particles in the population and called *global best* (*gbest*) position. Without loss of generality, if we consider the following minimization problem:

$$\text{Minimize} \quad f(X), s.t.\, X \in S \subseteq R^D, \tag{4}$$

where $f(X)$ is an objective function and S is the feasible space, then $P_{i,k}$ ∓ can be updated by

$$P_{i,k} = \begin{cases} X_{i,k} & \text{if} \quad f(X_{i,k}) < f(P_{i,k-1}) \\ P_{i,k-1} & \text{if} \quad f(X_{i,k}) \geq f(P_{i,k-1}) \end{cases}. \tag{5}$$

G_k can be found by

$$G_k = P_{g,k}, \tag{6}$$

where

$$g = \arg \min_{1 \leq i \leq m} [f(P_{i,k})]. \tag{7}$$

The parameters $r_{i,k}^j$ and $R_{i,k}^j$ are sequences of two different random numbers distributed uniformly within (0, 1), which is denoted by $r_{i,k}^j, R_{i,k}^j \sim U(0,1)$. Generally, the value of $V_{i,k}^j$ is restricted in the interval $[-V_{\max}, V_{\max}]$.

2.2.2 The QPSO Algorithm

The trajectory analysis [30] demonstrated the fact that the convergence of the PSO algorithm may be achieved if each particle converges to its local focus, $p_{i,k} = (p_{i,k}^1, p_{i,k}^2, \ldots p_{i,k}^D)$ defined at the coordinates

$$p_{i,k}^{j} = \varphi_{i,k}^{j} \cdot P_{i,k}^{j} + (1 - \varphi_{i,k}^{j}) \cdot G_{k}^{j}, \tag{8}$$

where

$$\varphi_{i,k}^{j} = c_1 r_{i,k}^{j} \Big/ (c_1 r_{i,k}^{j} + c_2 R_{i,k}^{j}), \tag{9}$$

with regard to the random numbers $r_{i,k}^{j}$ and $R_{i,k}^{j}$ in (2). In PSO, the acceleration coefficients c_1 and c_2 are generally set to be equal, i.e., $c_1 = c_2$, and thus $\varphi_{i,k}^{j}$ is a sequence of uniformly distributed random numbers in (0, 1), i.e., $\varphi_{i,k}^{j} \sim U(0,1)$. Thus the Eq. (8) can be restated as

$$p_{i,k}^{j} = \varphi_{i,k}^{j} \cdot P_{i,k}^{j} + (1 - \varphi_{i,k}^{j}) \cdot G_{k}^{j}, \ \varphi_{i,k}^{j} \sim U(0,1). \tag{10}$$

In QPSO [18, 19], each single particle is treated as a spin-less one moving in quantum space. Thus, the state of the particle is characterized by a wave function ψ, where $|\psi|^2$ is the probability density function of its position. Inspired by the convergence analysis of the particle in PSO, we assume that, at the kth iteration, particle i flies in the D-dimensional quantum space with a δ potential well centered at $p_{i,k}^{j}$ on the jth dimension ($1 \le j \le D$).

Let $Y_{i,k+1}^{j} = |X_{i,k+1}^{j} - p_{i,k}^{j}|$, we can obtain the normalized wave function at iteration $k + 1$

$$\psi(Y_{i,k+1}^{j}) = \frac{1}{\sqrt{L_{i,k}^{j}}} \exp\left(\frac{-Y_{i,k+1}^{j}}{L_{i,k}^{j}}\right), \tag{11}$$

which satisfies the bound condition that

$$\psi(Y_{i,k+1}^{j}) \to 0 \quad \text{as} \quad Y_{i,k+1}^{j} \to \infty. \tag{12}$$

$L_{i,k}^{j}$ is the characteristic length of the wave function. By the definition of the wave function, the probability density function is given by

$$Q(Y_{i,k+1}^{j}) = |\psi(Y_{i,k+1}^{j})|^2 = \frac{1}{L_{i,k}^{j}} \exp\left(\frac{-2Y_{i,k+1}^{j}}{L_{i,k}^{j}}\right), \tag{13}$$

and thus the probability distribution function is

$$F(Y_{i,k+1}^{j}) = 1 - \exp\left(\frac{-2Y_{i,k+1}^{j}}{L_{i,k}^{j}}\right). \tag{14}$$

Using the Monte Carlo method, we can measure the jPthP component of position of particle i at the $(k + 1)$Pth iteration by

$$X_{i,k+1}^{j} = p_{i,k}^{j} \pm \frac{L_{i,k}^{j}}{2} \ln\left(\frac{1}{u_{i,k+1}^{j}}\right), \quad u_{i,k+1}^{j} \sim U(0,1), \tag{15}$$

where $u_{i,k+1}^{j}$ is a sequence of random numbers uniformly distributed on (0, 1).

In [18], the value of $L_{i,k}^{j}$ is determined by

$$L_{i,k}^{j} = 2\alpha |X_{i,k}^{j} - p_{i,k}^{j}|. \tag{16}$$

Accordingly, Eq. (15) becomes

$$X_{i,k+1}^{j} = p_{i,k}^{j} \pm \alpha |X_{i,k}^{j} - p_{i,k}^{j}| \ln\left(\frac{1}{u_{i,k+1}^{j}}\right), \quad u_{i,k+1}^{j} \sim U(0,1). \tag{17}$$

In [19], $L_{i,k}^{j}$ is suggested to be determined by

$$L_{i,k}^{j} = 2\alpha |X_{i,k}^{j} - C_{k}^{j}|. \tag{18}$$

where $C_k = (C_k^1, C_k^2, \cdots, C_k^D)$ is called the mean best (*mbest*) position, defined by the average of the *pbest* positions of all particles [19], that is,

$$C_k^j = \frac{1}{m}\sum_{i=1}^{m} P_{i,k}^{j}, \quad (1 \le j \le D). \tag{19}$$

Therefore, the position of the particle updates according to the following equation:

$$X_{i,k+1}^{j} = p_{i,k}^{j} \pm \alpha |X_{i,k}^{j} - C_{k}^{j}| \ln\left(\frac{1}{u_{i,k+1}^{j}}\right), \quad u_{i,k+1}^{j} \sim U(0,1). \tag{20}$$

The parameter α in Eqs. (16), (17), (18) and (20) is called contraction-expansion (CE) coefficient, which can be adjusted to balance the local and global search of the algorithm during the optimization process. The PSO with Eq. (17) or (20) is called quantum-behaved particle swarm optimization (QPSO). To distinguish them, we denote the QPSO with (17) as QPSO-Type 1 and that with (20) as QPSO-Type 2.

The procedure of the QPSO is similar to that of the PSO algorithm, except that they have different evolution equations. In the QPSO algorithm, there is no velocity vector for each particle and the position of the particle updates directly according to equation either (17) or (20). For ease of implementation, here is a description on the flow of the procedure. The QPSO algorithm starts with the initialization of the current positions of the particles and their personal best positions by setting $P_{i,0} = X_{i,0}$, follows with the iterative update of the particle swarm using the update equations provided above. In each iteration of the procedure, the mean best position of the particle swarm is computed (for QPSO-Type 2) and the current position of each particle is updated according to Eq. (17)

or (20) with the coordinates of its local focus evaluated by Eq. (10). After each particle updates its current position, its fitness value is evaluated together with an update of the personal best position and the current global best position. In Eq. (17) or (20), the probability of using either "+" operation or "−" operation is equal to 0.5. The iteration continues until the termination condition is met.

The procedure of the QPSO algorithm is outlined in Fig. 3. Note that rand$i(\cdot)$, $i = 1, 2, 3$, is used to denote random numbers generated uniformly and distributed on (0,1). When the procedure is used for solving an optimization problem, the value of the CE coefficient α must be determined.

In [40], the influence of the CE coefficient α on the dynamical behaviour of an individual particle in the QPSO was analyzed theoretically. It was shown that,

Procedure of the QPSO algorithm:
Begin
Initialize the current position $X_{i,0}^j$ and the personal best position $P_{i,0}^j$ of each particle, evaluate their fitness values and find the global best position G_0 ; Set k=0.
While (termination condition = false)
Do
Set k=k+1;
Compute mean best position C_k (for QPSO-Type 2);
Select a suitable value for α ;
for (i=1 to m)
for j=1 to D
$\varphi_{i,k}^j = \text{rand1}(\cdot)$;
$p_{i,k}^j = \varphi_{i,k}^j P_{i,k}^j + (1-\varphi_{i,k}^j) G_k^j$;
$u_{i,k+1}^j = \text{rand2}(\cdot)$;
if ($\text{rand3}(\cdot) < 0.5$)
$X_{i,k+1}^j = p_{i,k}^j + \alpha \mid X_{i,k}^j - p_{i,k}^j \mid \ln(1/u_{i,k+1}^j)$ (for QPSO-Type 1);
(or $X_{i,k+1}^j = p_{i,k}^j + \alpha \mid X_{i,k}^j - C_k^j \mid \ln(1/u_{i,k+1}^j)$ (for QPSO-Type 2));
else
$X_{i,k+1}^j = p_{i,k}^j - \alpha \mid X_{i,k}^j - p_{i,k}^j \mid \ln(1/u_{i,k+1}^j)$ (for QPSO-Type 1);
(or $X_{i,k+1}^j = p_{i,k}^j - \alpha \mid X_{i,k}^j - C_k^j \mid \ln(1/u_{i,k+1}^j)$ (for QPSO-Type 2));
end if
end for
Evaluate the fitness value of $X_{i,k+1}$, i.e. the objective function value $f(X_{i,k+1})$;
Update $P_{i,k}$ and G_k ;
end for
end do
end

Fig. 3 The procedure of the QPSO algorithm

when $\alpha > e^{\gamma} \approx 1.781$, where $\gamma \approx 0.577215665$ is called the Euler constant, the particle diverges and the whole particle swarm explodes [40]. The QPSO-Type 2 has been shown to have a better overall performance than the QPSO-Type 1 [40–42]. Therefore, in this paper, the version of the QPSO used is the QPSO-Type 2, where the CE coefficient is controlled to decrease linearly from 1.0 to 0.5, as recommended in [40].

3 The GP-QPSO Method

The GP-QPSO method for modeling and optimization of a fermentation process contains three execution steps. The first step is to use the GP algorithm to model the relationship between the independent and dependent variables of the fermentation process with the given experimental data. The terminal nodes for the GP are the independent variables. For example, when modeling the HA production process, the terminal nodes are the culture conditions, including agitation speed (X_1), aeration rate (X_2) and stirrer number (X_3). In order to determine the function nodes, we should take into account the physical meaning of the fermentation process. Generally speaking, only some simple operators are suitable, such as "+", "−", "*", "/", "square", and "cube". The periodical functions, including "sin" and "cos" are undesirable since fermentation is an asymptotic process, not a periodical one. The fitness function for the GP in the proposed approach is the error function given by

$$e(i,t) = \sum_{j=1}^{N_c} \left(s(i,j) - c(j)\right)^2, \tag{21}$$

where $s(i,j)$ is the value of dependent variable computed using the expression represented by the ith individual on the values of the independent variables in the jth experiment; $c(j)$ means the value of the dependent variable in the jth experiment, and N_c denotes the number of the experiments. During the evolution of the GP, the parameters of the model (i.e., the mathematical expression given an individual) are randomly selected at given interval.

The second step for the GP-QPSO method is to optimize the parameters of the model (i.e., the factors of the variables) obtained by the GP algorithm. Here, the dimension of the search space depends on the number of the parameters of the obtained model. It should be noted that the number of the parameters varies in each different run of the GP algorithm, which may generate a completely different model. The fitness function in this step is the one given by (21). Here, the version of the QPSO used for optimizing the parameters of the model is the QPSO-Type 2. The procedure of this step for the GP-QPSO method is visualized in the flow chart in Fig. 4.

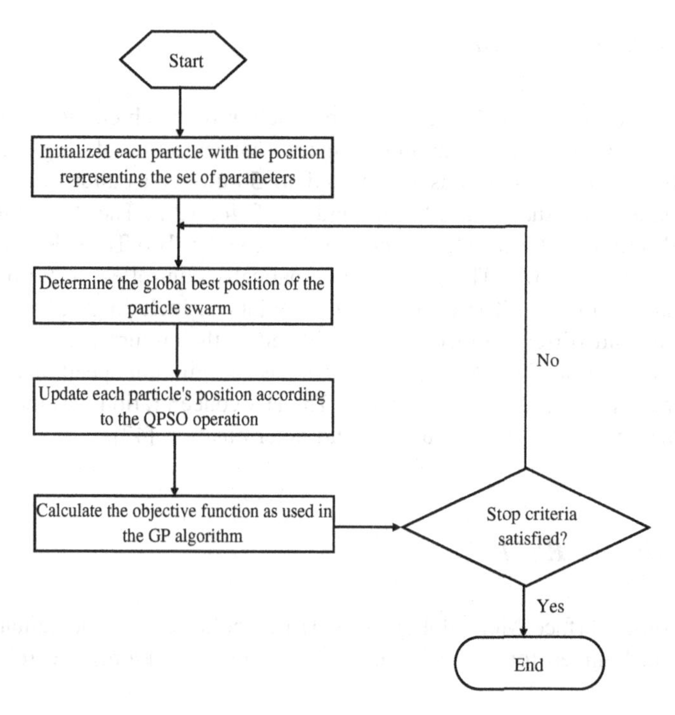

Fig. 4 The procedure of the second step for the GP-QPSO method

The final step is to optimize the values of the independent variables by using the QPSO in order to maximize the dependent variable, which generally represents the concentration of a fermentation product. The purpose of this optimization task is to increase the production of the fermentation process by optimizing the culture conditions. The dimension of the search space in this step is the number of the dependent variables. The mathematical expression of the model now serves as the fitness function. The final global best position found by the QPSO is thus the optimized values of the independent variables, and its fitness value is the maximized value of the dependent variable, i.e., the production of the fermentation product.

4 Experimental Results

The proposed GP-QPSO approach was applied to the HA production process. A performance comparison was also made between the RSM and the GP-QPSO method.

4.1 Experimental Data

To acquire real experimental data, a HA production by batch culture of *S. zooepidemicus* was carried out with an initial sucrose concentration of 70 g/l. One loop of cells from a fresh slant was transferred to 50 ml seed culture medium and cultured on a rotary shaker at 200 rpm and 37 °C for 12 h. The seed culture was inoculated into a 7-l fermentor (Model KL-7l, K3T Ko Bio Tech, Korea) with a working volume of 4.0 l. The pH was automatically controlled at 7.0 by adding 5 mol/l NaOH solution. Three independent variables, including agitation speed, aeration rate and stirrer number were considered as the culture conditions for the process optimization. The detailed level designs of agitation speed, aeration rate and stirrer number were listed in Table 1. The HA concentration was measured by the carbazole method based on uronic acid determination [43].

4.2 Results for RSM

The Response Surface Methodology (RSM) is a collection of mathematical and statistical techniques for empirical model building. It explores the relationship

Table 1 The Box-Behnken experimental design with three independent variables for the HA production

Experiments	Agitation speed (rpm)		Aeration rate (vvm)		Stirrer number		HA yield (g/l)
	X_1	Coding X_1	X_2	Coding X_2	X_3	Coding X_3	Y
1	200	−1	2.0	1	3	−1	4.9
2	300	0	1.5	0	4	0	5.2
3	400	1	2.0	1	5	1	4.1
4	200	−1	1.0	−1	3	−1	4.5
5	300	0	1.5	0	2	−1.682	4.3
6	400	1	1.0	−1	3	−1	4.3
7	100	−1.682	1.5	0	4	0	3.6
8	300	0	1.5	0	4	0	5.2
9	300	0	1.5	0	4	0	5.2
10	300	0	1.5	0	4	0	5.2
11	200	−1	1.0	−1	5	1	4.5
12	400	1	2.0	1	3	−1	4.7
13	500	1.682	1.5	0	4	0	3.9
14	200	−1	2.0	1	5	1	4.0
15	400	1	1.0	−1	5	1	4.6
16	300	0	1.5	0	4	0	5.2
17	300	0	1.5	0	4	0	5.2
18	300	0	1.5	0	6	1.682	4.8
19	300	0	0.5	−1.682	4	0	4.7
20	300	0	2.5	1.682	4	0	5.1

between several explanatory variables (i.e., independent variables or inputs) and one or more response variables (i.e., dependent variables). In general, such a relationship is unknown but can be approximated by a low-degree polynomial form. Two important models are commonly used in RSM: one is the first-order model,

$$y = \beta_0 + \sum_{i=1}^{N} \beta_i x_i + \varepsilon, \tag{22}$$

and the other is the second-order model,

$$y = \beta_0 + \sum_{i=1}^{N} \beta_i x_i + \sum_{i<j} \beta_{ij} x_i x_j + \sum_{i=1}^{N} \beta_{ii} x_i^2 + \varepsilon, \tag{23}$$

where β_0, β_i, $(1 \leq i \leq N)$, $\beta_{ij}(1 < i < j,\ 1 \leq i \leq N)$, and $\beta_{ii}(1 \leq i \leq N)$ are unknown coefficient constants referred to as parameters, N denotes the number of independent variables, and ε is a random experimental error assumed to have a zero mean. The most widely used is the second-order model, which was also employed in this study.

By applying multiple regression analysis on the experimental data, the following second-order polynomial equation was developed, which identifies the relationship between the HA production (Y) and the agitation speed (X_1), aeration rate (X_2) and stirrer number (X_3):

$$\begin{aligned} Y = {} & 5.1968 + 0.0223X_1 + 0.0346X_2 + 0.0263X_3 - 0.4918X_1^2 \\ & - 0.0853X_2^2 - 0.2090X_3^2 - 0.0012X_1X_2 - 0.0750X_1X_3 - 0.225X_2X_3 \end{aligned} \tag{24}$$

By calculating the derivative of the function in the right side of Eq. (24), the optimal values of X_1, X_2 and X_3 (in the coded units) were found to be -1.4, -0.71 and -1.00, respectively. Correspondingly, we can obtain the maximum point of the model, which was 260 rpm for agitation speed, 1.15 vvm for aeration rate, and 3 for stirrer number. The maximum predicted value of the HA production (Y) is 5.27 g/l. It should be noted that β_0 is a large coefficient. In fact, the microbial HA production of *S. zooepidemicus* can be conducted in aerobic or anaerobic conditions. Usually, the aerobic culture of *S. zooepidemicus* can produce more HA than the anaerobic culture due to the increased energy status of the cells. The factors influencing the HA production include not only the mixing performance, but also the metabolic efficiency of the cells. Therefore, the large coefficient for β_0 is possibly due to the contribution of the other factors except the mixing performance considered in this work. Table 2 lists the results of the estimated parameters and P-values, which suggested that the second-order model of RSM fits well the HA fermentation process.

Table 2 The results of regression analysis

Parameters	Values	*P* value
β_0	5.1968	0.01
β_1	0.0223	0.02
β_2	0.0346	0.12
β_3	0.0263	0.23
β_{11}	−0.4918	0.01
β_{22}	−0.0853	0.06
β_{33}	−0.2090	0.09
β_{12}	−0.0012	0.02
β_{13}	−0.0750	0.01
β_{23}	−0.225	0.13

4.3 Results for Modeling with GP

When the GP algorithm was used for modeling the fermentation process of the HA production, the terminal nodes of an individual represented the agitation speed (X_1), aeration rate (X_2) and stirrer number (X_3); the set of the function nodes was (+, −, *, /, square, cube); the population size was 40; the crossover rate was 0.8; the mutation rate was 0.2; and the number of the maximum generation is 100.

The GP algorithm was run for 30 times and the resulting model with the best fitness value 0.6325 is given by

$$\begin{aligned} Y = 5.1663 - 0.0179X_1 + 0.0774X_2 - 0.2346X_3 - 0.4844X_1^2 \\ - 0.0985X_2^2 - 0.2008X_3^2 + 0.0220X_1^3 - 0.0048X_2^3 + 0.1323X_3^3 \end{aligned} \tag{25}$$

The convergence process the GP algorithm is visualized in Fig. 5, which shows that the fitness value, namely the error converged to the best fitness value after 70 generations.

4.4 Results for Modeling with the GP-QPSO Method

In the modeling of the fermentation process of the HA production with the GP-QPSO method, the QPSO algorithm was used to optimize the parameters of the model generated by the GP algorithm as described in Eq. (25). It can be seen that the number of the parameters of the model is 10. Therefore the dimension of the search space is 10 and the current and *pbest* positions of the *i*th particle at the *k*th iteration of the QPSO are denoted as $X_{i,k} = (X_{i,k}^1, X_{i,k}^2, \ldots, X_{i,k}^{10})$ and $P_{i,k} = (P_{i,k}^1, P_{i,k}^2, \ldots, P_{i,k}^{10})$ respectively, with each component of the position vector representing a parameter (i.e., a coefficient) of the model. The objective function (i.e., the fitness function) is still the error function described by Eq. (21).

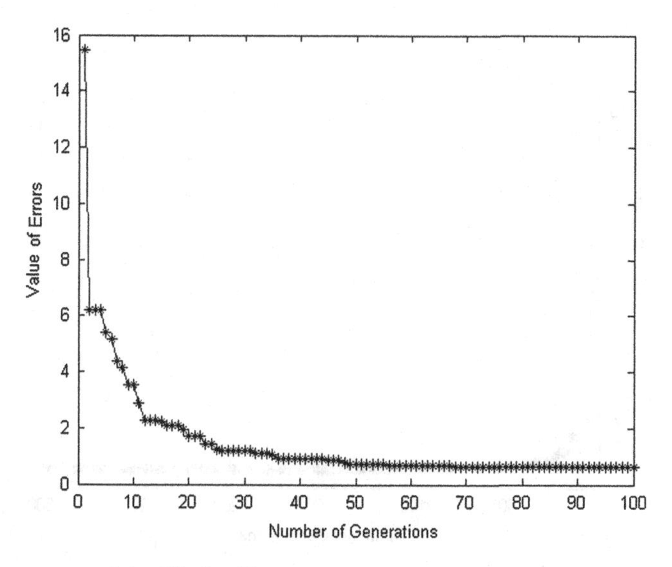

Fig. 5 Convergence of the GP algorithm

When employing the QPSO algorithm for the optimization task, we used 40 particles for the algorithm, which ran for 30 times, with each run executed for 300 iterations. The CE coefficient α of the QPSO decreased linearly from 1.0 to 0.5 on the course of the running. The resulting model with the best fitness value 0.4809 out of 10 runs of the QPSO is described by

$$Y = 5.1968 - 0.0874X_1 - 0.1037X_2 - 0.3133X_3 - 0.4918X_1^2 - 0.0853X_2^2 - 0.2090X_3^2 + 0.0624X_1^3 + 0.0787X_2^3 + 0.1633X_3^3 \quad (26)$$

The convergence process of the GP-QPSO modeling is traced in Fig. 6. It can be observed that the algorithm had a good convergence, with the fitness value converging to the best value 0.4809 after almost 130 iterations.

Table 3 compares the performance of the RSM, GP and GP-QPSO by listing their fitness values, i.e., the errors of the models on the given experimental data. Table 4 provides the results of t test between the means of the errors of the models with 0.05 as the level of significance. As evident from the results, the GP-QPSO and the GP methods generated models with better qualities than the RSM, for their errors are the significantly lower than the RSM model. Between the GP-QPSO and the GP methods, the mean error of the models generated by the former method showed to be significantly better than the models by the latter one. This implies that using the QPSO to tune the parameters of the model obtained by the GP algorithm can improve the fitting performance of the model.

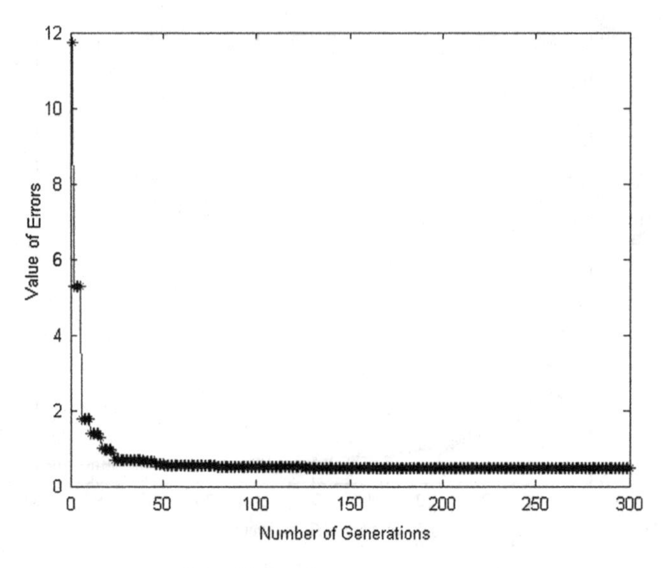

Fig. 6 Convergence process of the GP-QPSO

Table 3 The comparison between the results of the GP, the GP-QPSO and the RSM

Modeling methods	Minimum final errors	Mean final errors	Standard deviation of the errors
RSM	0.6560	0.6560	–
GP	0.6325	0.6389	0.0014
GP-QPSO	0.4809	0.4812	0.0001

Table 4 The results of *t* test between the mean final errors of the models obtained by the three modeling methods

	GP-QPSO versus RSM	GP-QPSO versus GP	GP versus RSM
Standard error	1.8257e-005	2.5626e-004	2.5560e-004
95 % Confidence interval	(−0.17484, −0.17477)	(−0.1582, −0.1572)	(−0.0176, −0.0166)
T-value	9574.1903	615.4024	66.9004
P value	<0.0001	<0.0001	<0.0001
Significance	Extremely significant	Extremely significant	Extremely significant

4.5 Results on the Optimization of the Culture Conditions

Based on the model produced by the GP-QPSO modeling method, we further optimize with the QPSO algorithm the independent variables, namely the culture conditions including the agitation speed (X_1), the aeration rate (X_2) and the stirrer number (X_3), in order to maximize the HA production. In this optimization procedure, the dimension of the search space is 3, the number of the culture conditions. For particle i at the kth iteration, its current position can be denoted as

$X_{i,k} = (X_{i,k}^1, X_{i,k}^2, X_{i,k}^3)$ and its personal best position as $P_{i,k} = (P_{i,k}^1, P_{i,k}^2, P_{i,k}^3)$. The objective function (i.e., the fitness function) in this optimization task is the model itself, i.e., the function expressed by Eq. (26). In the experiment, twenty particles were used for the optimization task and the CE coefficient decreased linearly from 1.0 to 0.5 over the running of the algorithm. The QPSO ran 10 times, with each run lasting 100 iterations.

The best fitness value, i.e., the best HA yield obtained out of 10 runs of the QPSO algorithm, was 5.57 g/l, with the optimized agitation speed 278 rpm, aeration rate 1.65 vvm and stirrer number 3, as shown in Table 5. The optimized values of the independent variables of the model by the RSM are also provided in the table. It can be found that the optimization of the GP-QPSO based model resulted in higher HA yield. The convergence process of the QPSO in this search is shown in Fig. 7. It can be seen that the fitness value increased very fast during the early stage of the search, converging to the optimal value after 10 iterations. Since this optimization problem is a polynomial one with dimensionality 3, it is an easy problem, so that the QPSO can find out its optimal solution rapidly.

Table 5 Culture conditions and HA yield, optimized for the models obtained by the RSM and GP-QPSO methods

Models	Agitation speed	Aeration rate	Stirrer number	HA yield (g/l)
RSM	260	1.15	3	5.27
GP-QPSO	278	1.65	3	5.57

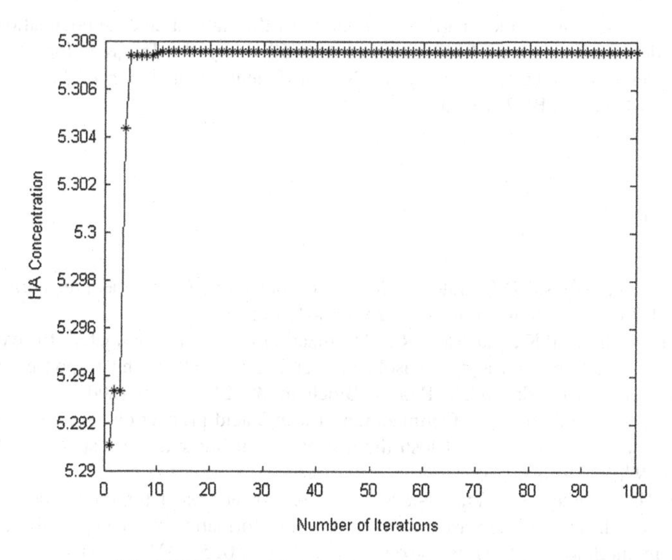

Fig. 7 Convergence of the QPSO algorithm in the optimization of the culture conditions

We tested the optimized culture conditions of the model by using the GP-QPSO method with the same experimental conditions as those for acquiring the experimental data. We found that the HA yield was 5.55, with the agitation speed 278 rpm, aeration rate 1.65 vvm and 3 stirrer, which indicates that the model generated by the GP-QPSO fits well the fermentation process of the HA production.

5 Conclusion

In this paper, we proposed a GP-QPSO method for the modeling and optimization of a fermentation process. This method firstly uses the GP algorithm to model the relationship between the independent variables and the dependent variable of the process. During the evolution of the GP, the parameters of the model represented by an individual are randomly selected within a given interval. The obtained model outputted by the GP is further improved by tuning its parameters with the QPSO algorithm. The independent variables of the resulting model, which represent the culture conditions of the fermentation process, are then optimized by the QPSO such that the dependent variable (i.e., the yield of the fermentation product) can be maximized.

The GP-QPSO approach was applied to the fermentation process of the HA production. With the given experimental data, the GP-QPSO was shown to fit the process better than the RSM method and the GP algorithm. It was also found that the optimization of the culture conditions of the model generated by the proposed method can lead to a better HA yield, which was verified by real experiments.

Acknowledgments This work is partially supported by the Natural Science Foundation of China (NSFC), under grant number 601190117 and 60975080, by the Program for New Century Excellent Talents in University, and by the Natural Science Foundation of Jiangsu Province, China (Project Number: BK2010143).

References

1. Kennedy, M., Krouse, D.: Strategies for improving fermentation medium performance: a review. J. Ind. Microbiol. Biotechnol. **23**, 456–475 (1999)
2. Dutta, J.R., Dutta, P.K., Banerjee, R.: Optimization of culture parameters for extracellular protease production from a newly isolated pseudomonas SP. Using response surface and artificial neural network models. Process Biochem. **39**, 2193–2198 (2004)
3. Sim, J.H., Kamaruddin, A.H.: Optimization of acetic acid production from synthesis gas by chemolithotrophic bacterium—Clostridium aceticum using statistical approach. Bioresour. Technol. **99**, 2724–2735 (2008)
4. Ceylan, H., Kubilay, S., Aktas, N., Sahiner, N.: An approach for prediction of optimum reaction conditions for laccase-catalyzed bio-transformation of 1-naphthol by response surface methodology (RSM). Bioresour. Technol. **99**, 2025–2031 (2008)

5. Chang, S.W., Shaw, J.F., Yang, K.H., Chang, S.F., Shieh, C.J.: Studies of optimum conditions for covalent immobilization of Candida rugosa lipase on poly (Gamma-Glutamic Acid) by RSM. Bioresour. Technol. **99**, 2800–2805 (2008)
6. Kunamneni, A., Singh, S.: Response surface optimization of enzymatic hydrolysis of maize starch for higher glucose production. Biochem. Eng. J. **27**, 179–190 (2005)
7. Ustok, F.I., Tari, C., Gogus, N.: Solid-state production of polygalacturonase by Aspergillus sojae ATCC 20235. J. Biotechnol. **127**, 322–334 (2007)
8. Cramer, N.L.: A representation for the adaptive generation of simple sequential programs. In: Grefenstette, J.J. (ed.) Proceedings of an International Conference on Genetic Algorithms and the Applications. Carnegie Mellon University (1985)
9. Koza, J.R.: Genetic programming: a paradigm for genetically breeding populations of computer programs to solve problems. Stanford University Computer Science Department, technical report STAN-CS-90-1314 (1990)
10. Koza, J.R.: Genetic Programming: On the Programming of Computers by Means of Natural Selection. MIT Press, Cambridge (1992)
11. Koza, J.R.: Genetic Programming II: Automatic Discovery of Reusable Programs. MIT Press, Cambridge (1994)
12. Koza, J.R., Bennett, F.H., Andre, D., Keane, M.A.: Genetic Programming III: Darwinian Invention and Problem Solving. Morgan Kaufmann, San Francisco (1999)
13. Koza, J.R., Keane, M.A., Streeter, M.J., Mydlowec, W., Yu, J., Lanza, G.: Genetic Programming IV: Routine Human-Competitive Machine Intelligence. Kluwer Academic Publishers, Boston (2003)
14. Mckay, B., Chen, S.-H., Nguyen, X.H.: Genetic programming: an emerging engineering tool. Int. J. Knowl Based Intell. Eng. Syst. **12**(1), 1–2 (2008)
15. Korns, M.: Large-scale, time-constrained, symbolic regression-classification. In: Genetic Programming Theory and Practice V. Springer, New York (2007)
16. Korns, M.: Symbolic regression of conditional target expressions. In: Genetic Programming Theory and Practice VII. Springer, New York (2009)
17. Korns, M.: Abstract expression grammar symbolic regression. In: Genetic Programming Theory and Practice VIII. Springer, New York (2010)
18. Sun, J., Feng, B., Xu, W.-B.: Particle swarm optimization with particles having quantum behavior. In: Proceedings of the 2004 Congress on Evolutionary Computation, pp. 326–331. IEEE Press (2004)
19. Sun, J., Xu, W.-B., Feng, B.: A global search strategy of quantum-behaved particle swarm optimization. In: Proceedings of the 2004 IEEE Conference on Cybernetics and Intelligent Systems, pp. 111–116. IEEE Press (2004)
20. Kennedy, J.: Bare bones particle swarms. In: Proceedings of the 2003 IEEE Swarm Intelligence Symposium, pp. 80–87. IEEE Press, Indianapolis, IN, April 2003
21. Kennedy, J.: Probability and dynamics in the particle swarm. In: Proceedings of the 2004 Congress on Evolutionary Computation, vol. 1, pp. 340–347. IEEE Press, June 2004
22. Fang, W., Sun, J., Ding, Y., Wu, X., Xu, W.: A review of quantum-behaved particle swarm optimization. IETE Tech. Rev. **27**, 336–348 (2010)
23. Banzhaf, W., Nordin, P., Keller, R.E., Francone, F.D.: Genetic Programming: An Introduction. Morgan Kaufmann, San Francisco (1998)
24. Koza, J.R.: Genetic Programming, vol. I. MIT Press, New York (1992)
25. Langdon, W.B.: Data Structures and Genetic Programming, Advances in Genetic Programming 2. MIT Press, Cambridge (1996)
26. Kennedy, J. Eberhart, R.C.: Particle swarm optimization. In: Proceedings of the 1995 IEEE International Conference on Neural Networks, pp. 1942–1948. Piscataway, NJ (1995)
27. Jones, K.O.: Comparison of genetic algorithm and particle swarm optimization. In: Proceedings of the 2005 International Conference on Computer System and Technologies, pp. IIIA1-6 (2005)
28. Kennedy, J.: The behavior of particle. In: Proceedings of the 7th Annual Conference on Evolutionary Programming, pp. 581–589 (1998)

29. Ozcan, E., Mohan, C.K.: Particle swam optimization: surfing the waves. In: Proceedings of the 1999 IEEE Congress on Evolutionary Computation, pp. 1939–1944 (1999)
30. Clerc, M., Kennedy, J.: The particle swarm-explosion, stability and convergence in a multidimensional complex space. IEEE Trans. Evol. Comput. **6**(2), 58–73 (2002)
31. Kadirkamanathan, V., Selvarajah, K., Fleming, P.J.: Stability analysis of the particle dynamics in particle swarm optimizer. IEEE Trans. Evol. Comput. **10**(3), 245–255 (2006)
32. Shi, Y., Eberhart, R.C.: A modified particle swarm optimizer. In: Proceedings of the IEEE International Conference on Evolutionary Computation, pp. 69–73 (1998)
33. Clerc, M.: The swarm and the queen: towards a deterministic and adaptive particle swarm optimization. In: Proceedings of the 1999 Congress on Evolutionary Computation, vol. 3, pp. 1951–1957 (1999)
34. Bratton, D., Kennedy, J.: Defining a standard for particle swarm optimization. In: Proceedings of the 2007 IEEE Swarm Intelligence Symposium, pp. 120–127 (2007)
35. Suganthan, P.N.: Particle warm optimizer with neighborhood operator. In: Proceedings of the 1999 Congress on Evolutionary Computation, pp. 1958–1961 (1999)
36. Liang, J.J., Suganthan, P.N.: Dynamic multiswarm particle swarm optimizer (DMS-PSO). In: Proceedings of the 2005 IEEE Swarm Intelligence Symposium, pp. 124–129 (2005)
37. Mendes, R., Kennedy, J., Neves, J.: The fully informed particle swarm: simpler, maybe better. IEEE Trans. Evol. Comput. **8**(3), 204–210 (2004)
38. Van den Bergh, F., Engelbrecht, A.P.: A cooperative approach to particle swarm optimization. IEEE Trans. Evol. Comput. **8**(3), 225–239 (2004)
39. Poli, R.: Analysis of the publications on the applications of particle swarm optimisation. J. Artif. Evol. Appl. **2008**, 1–10 (2008)
40. Sun, J., Fang, W., Wu, X., Palade, V., Xu, W.: Quantum-behaved particle swarm optimization: analysis of the individual particle behavior and parameter selection. Evol. Comput. **20**(3), 349–393 (2012)
41. Sun, J., Wu, X., Palade, V., Fang, W., Lai, C.-H., Xu, W.: Convergence analysis and improvements of quantum-behaved particle swarm optimization. Inf. Sci. **193**, 81–103 (2012)
42. Sun, J., Fang, W., Palade, V., Wu, X., Xu, W.: Quantum-behaved particle swarm optimization with Gaussian distributed local attractor point. Appl. Math. Comput. **218**(7), 3763–3775 (2011)
43. Bitter, H., Muir, H.M.: A modified uronic acid carbazole reaction. Anal. Biochem. **4**, 330–334 (1962)

Hybrid Client Specific Discriminant Analysis and its Application to Face Verification

Xiao-Qi Sun, Xiao-Jun Wu, Jun Sun and Philippe Montesinos

Abstract Techniques to perform dimensionality reduction for high dimensional data can vary considerably from each other, leading to different effects on face verification. To address the problem, we introduce a framework called the hybrid client specific discriminant analysis that incorporates various dimensionality reduction methods with client specific subspace. In contrast to the common multidimensional representation, like PCA and LDA, client specific subspace could better describe the diversity of the different faces and has more robust discriminatory information. Moreover, it provides two measures for authentication: a distance to the client template and a distance to the mean of imposter. These two decision scores are combined to achieve significant performance gains. Extensive experiments obtained on the facial databases XM2VTS show the effectiveness of the hybrid client specific discriminant analysis.

Keywords Face verification · Dimensionality reduction · Client specific · Hybrid client specific discriminant analysis

1 Introduction

Human face recognition is one of the most valuable biometric identification methods among different modalities of biometrics, which involves image processing, pattern recognition, computer vision and others. It has been a hot topic

X.-Q. Sun · X.-J. Wu (✉) · J. Sun
School of IoT Engineering, Jiangnan University, 214122 Wuxi, China
e-mail: Wu_xiaojun@yahoo.com.cn

P. Montesinos
LGI2P, Ecole De Mines D'Ales, Parc Scientifique Georges Besse, 30035 NIMES cedex 1, France
e-mail: Philippe.Montesinos@mines-ales.fr

I. Hatzilygeroudis and V. Palade (eds.), *Combinations of Intelligent Methods and Applications*, Smart Innovation, Systems and Technologies 23,
DOI: 10.1007/978-3-642-36651-2_8, © Springer-Verlag Berlin Heidelberg 2013

over the past years. In terms of application, face recognition has two types; one of them is human identity recognition, which recognizes the identity of the person with face images and aims to solve "who are you"; the other is face verification, which determines whether the one is the claimed person and aims to solve whether you are the one [1].

Over the past decades, the study of personal identity verification is an attractive topic in the field of pattern recognition and has received considerable attention. A large number of approaches have been proposed in the literature [2–6]. Feature extraction is the most fundamental problem in face verification. The goal of feature extraction is to project the original high dimensional data onto a lower dimensional feature space [7]. For pattern classification, the purpose of feature extraction is to map the original data onto a discriminative feature space in which the samples from different classes are clearly separated [8–11]. However, techniques to perform dimensionality reduction (DR) for high dimensional data can vary considerably from each other. Overall, we categorize them as follows: global structure DR, local structure DR and Kernel method.

1.1 Global Structure Dimensionality Reduction

The most commonly used global representation for face recognition and verification is the Karhunen–Loeve (KL) expansion [12] which is also known as the Principal Component Analysis (PCA). Its application to the face recognition problem has been pioneered by Sirovich and Kirby [13, 14], but the approach has been popularized by Turk and Pentland [15] where the PCA bases are referred to as eigenfaces. Since then, numerous different face recognition approaches have been developed and successfully applied in the real word. Examples can be found in [16–19]. A detailed analysis of PCA-based face recognition algorithms can be found in [20].

The PCA technique yields projection directions that maximize the total scatter across all classes, which does not guarantee the most efficient compression of discriminatory information because there is no category information which is embedded in the total scatter matrix. The linear discriminant analysis (LDA) [12, 21] has been adapted to face recognition by Belhumeur et al. [22]. The objective of LDA is to find a subspace that maximizes the distance from different classes and meanwhile minimizes the distance from the same class in the LDA subspace. Thus, the derived subspace is discriminative to classify different samples correctly. However, in real applications, due to the high dimensionality of feature and usually small number of samples, the classical LDA always fails because of the small sample size (SSS) problem. To address this problem, Swets [23] advocated that the face data should first be projected into a PCA space to ensure that the within-class matrix is not rank deficient. The LDA-based features are then extracted in this lower dimensional space. Face recognition or verification using Fisher faces was studied by many authors including [23–28], etc.

However, most of the above subspace learning methods like PCA, LDA and so on, consider the pixels in image independently, not taking into account their spatial relationship [29]. It is well known that images with certain pattern occupy specific manifold, which is constrained by contextual information in high-dimensional feature space. Therefore, contextual constraint in image is important for classification. One of the most successful works to model the contextual information is the Markov Random Fields (MRFs) [30–32] which derive the results by maximizing the posterior probability in Bayesian deduction framework. However, the optimization by MRF is somewhat computationally expensive and is easy to converge toward local minima, which limits its application. Wang et al. [33] proposed a novel image matching distance considering the spatial information. However, they did not demonstrate how to integrate the contextual information into dimensionality reduction problem. In [29], Lei and Li proposed contextual constraints-based linear discriminant analysis (CCLDA) which incorporates the contextual information into LDA. The main difference of CCLDA from the previous ones is that it takes into account the image contextual constraints during in dimensionality reduction.

1.2 Local Structure Dimensionality Reduction

As mentioned above, the conventional global feature extraction methods, like PCA, LDA, PCA + LDA, etc. are applied on the whole image to extract global discriminant facial features. The performance of these methods may degrade when local facial changes occurred due to e.g. variations of facial expression, illumination condition, pose, etc.

To address these problems, nonlinear methods by virtue of manifold learning have aroused broad interest in the few years. Contrasting with PCA and LDA, which aims to preserve the global Euclidean structure of the data space, manifold learning algorithms aim to preserve the inherent manifold structure [7]. Isometric Feature Mapping (ISOmap) [34], Locally Linear Embedding (LLE) [35], and Laplacian Eigenmap [36]. He proposed Local Preserving Projections (LPP) [37] which can find the optimal linear approximations to the Eigen-functions of the Laplace–Beltrami operator on the manifold. So LPP approach preserves local structure and seems to have more discriminant power than PCA. All of these algorithms attempt to embed the original data into a sub manifold by preserving the local neighborhood structure. They are based on the assumption that the manifold's intrinsic geometry can be fully determined by the local metric and neighborhood information. Different from LLE, ISOmap and Laplacian eigenmap, LPP is a linear algorithm. It is quite simple and easy to implement. [38].

1.3 The Kernel Method

The human face plays a major role in convey personal identity. However, both pose and illumination conditions will vary in less controlled scenarios, which make it impossible to solve the face verification problem using a linear method, because of the complex nature of the human face data. Since much of the important information may be contained in the high order relationships among the image pixels of a face pattern, the study of kernel method is an attractive topic in the field of pattern recognition over the last 10 years. Kernel PCA (KPCA) approach to face recognition was proposed in [39], which is a kernel version of Eigen face approach. However, KPCA does not exhibit satisfactory performance in face recognition, and nonlinear Fisher discriminant analysis using kernel technique was presented. To reduce the computational complexity, Yang proposed a method which is based on the combination of KPCA and LDA, in which the essence of kernel discriminant analysis is presented.

Despite great applicability, many dimensionality reduction methods often suffer two main restrictions. Firstly, many of them, especially the linear ones, require data to be represented in the form of feature vectors, which may eventually reduce the effectiveness of the overall algorithms when the data could be more precisely characterized in other forms. Secondly, there seems to lack a systematic way of integrating multiple image features for dimensionality reduction. To overcome the above-mentioned restrictions, Lin [40] introduced a framework called MKL-DR that incorporates multiple kernel learning (MKL) into training process of dimensionality reduction methods.

All the above approaches are based on a global representation of both the training samples and the probe in a subspace of the training data space, called feature space. To get better effectiveness in face verification, Kittler [41–44] proposed a client specific fisher face representation in which the client enrollment is insulated from the enrollment of other clients and there is only one fisher face per client. An improved model of client specific linear discriminant analysis (CSLDA) method was developed by Wu and Kittler [42]. In 2005, Wu and Kittler [45] further developed the nonlinear version of the improved CSLDA model based on kernel method, which is called client specific kernel discriminant analysis (CSKDA). Many attractive properties make the method ideally suited for both representation and authentication in personal identity verification systems and warrant its further development.

In this paper, a hybrid client specific discriminant analysis framework is studied. The proposed framework keeps the advantages of both client specific subspace and different dimensionality reduction techniques.

The rest of the paper is organized as follows. Several hybrid client specific discriminant analysis methods are developed in Sect. 2. Experimental results and analysis are presented in Sect. 3 and conclusions are drawn in Sect. 4.

2 Proposed Hybrid Client Specific Discriminant Analysis Methods

Since the relevant literature is quite extensive, our survey instead emphasizes the key concepts crucial to the establishment of the proposed methods. In order to present the proposed hybrid client specific discriminant analysis methods, we first review the theory of CSLDA in [42], then we develop hybrid client specific discriminant analysis methods by combining client specific technology with several other pattern recognition methods including CCLDA, LPP, and MKL.

2.1 The Theory of Client Specific Discriminant Analysis

The basic steps of client specific linear discriminant analysis (CSLDA) are as follows. First, the dimensionality reduction is performed on the original samples, then we construct a certain client model and its corresponding imposter model. Finally, a two-class LDA problem is developed to separate each clients and its corresponding imposters to obtain client specific Fisherface. To fully utilize the discriminant information provided by the training samples of CSLDA, we adopt an improved algorithm for client specific discriminant analysis proposed in [42].

A set of face images samples {zi} can be represented as an $K \times N$ matrix $Z = [z_1, z_2, \ldots, z_N]$, where K is the number of pixels in the images and N is the number of samples [46].

The between-class scatter matrix S_b is defined as $S_b = P_b P_b^T$, where $P_b = [\sqrt{N_1}(u_1 - u), \ldots, \sqrt{N_C}(u_C - u)]$, N_i is the number of samples of ith class ω_i, $u_i = \frac{1}{N_i}\sum_{j=1}^{N_i} z_j^i$, $z_j^i \in \omega_i$ is the mean vector of ith class and u is the total mean vector of all samples and C is the number of the clients. And assuming the population mean to be zero, it can be shown that the mean vector of imposter of ith class is $u_\Omega = -\frac{N_i}{N-N_i} u_i$.

To reduce the computational complexity, Turk and Pentland [15] suggested an indirect method to solve the eigenvectors of $S_b = P_b P_b^T$, which can be derived from the eigenvectors of the matrix $P_b^T P_b$. Let λ_i and e_i be the ith eigenvalue and its corresponding eigenvector of $P_b^T P_b$ Consequently, the optimal project vector $Y = [y_1, y_2, \ldots, y_d] = P_b E_d$, where $E_d = [e_1, e_2, \ldots, e_d]$ are the eigenvectors corresponding to the first d largest eigenvalues of $P_b^T P_b$.

Let $D_b = diag[\lambda_1, \lambda_2, \ldots, \lambda d]$, and further let $U = Y D_b^{-\frac{1}{2}}$. Projecting all the images into the subspace spanned by U, we have

$$\chi_i = U^T(z_i - u), \quad i = 1, 2, \ldots, N \tag{1}$$

Let us denote the mixture covariance matrix of the projected vectors by Φ, i.e.

$$\Phi = \frac{1}{N}\sum_{i=1}^{N} \chi_i \chi_i^T \tag{2}$$

Let us now consider the problem of discriminating class ω_i from all the other classes. In the context of the face verification problem this corresponds to discriminating between ith client and imposters modeled by all the other clients in the training data set. Given the mean vector of the ith class as:

$$\mu_i = \frac{1}{N_i}\sum_{i=1}^{N_i} \chi_i, \chi_i \in \omega_i \tag{3}$$

And assuming the population mean to be zero, it can be shown that the mean vector of imposters of ith class is:

$$\mu_\Omega = -\frac{N_i}{\mathrm{N} - \mathrm{N}_i}\mu_i \tag{4}$$

And the between-class scatter matrix is:

$$M_i = \frac{\mathrm{N}_i}{\mathrm{N} - N_i}\mu_i \mu_i^T \tag{5}$$

Then the within-class scatter matrix is:

$$\Sigma_i = \Phi - M_i \tag{6}$$

The Fisher discriminant function can be defined as:

$$J(v) = \frac{v^T M_i v}{v^T \sum_i v} \tag{7}$$

The solution to the problem can be found easily as:

$$v_i = \sum_i^{-1} \mu_i \tag{8}$$

Thus the overall client i specific discriminant transformation a_i :

$$a_i = U v_i \tag{9}$$

However, we can see from Eq. (7) that the inverse of the within-class scatter matrix should be calculated for each client. In the rest of the section, we utilize an equivalent Fisher criterion function proposed in [42]:

$$J(v) = \frac{v^T M v}{v^T \Phi v} \tag{10}$$

Similar to the above analysis, the optimal solution to the client specific discriminant problem can be found as:

$$\ddot{\upsilon} = \Phi^{-1}\mu_i \tag{11}$$

Similarly, the new client specific fisher face of the claimed identity can be given a:

$$a_i = U\ddot{\upsilon}_i \tag{12}$$

2.2 Contextual Constraints Based on Linear Discriminant Analysis

CCLDA incorporates the contextual information into linear discriminant analysis. In CCLDA, intuitively, if the pixels are of the similar property or reflect the similar structure, the weights on them would have strong relationship; otherwise the weights on independent pixels would also be weakly related. Following this idea, a constraint $J_2(w) = \frac{1}{2}\sum_{ij}(w_i - w_j)^2 S_{ij}$ is imposed on traditional LDA to formulate the objective of discriminant analysis as [29]:

$$J = \frac{w^T S_b w}{(w^T S_w w + \eta J_2(w))} \tag{13}$$

where

$$S_b = \frac{1}{N}\sum_{i=1}^{C} N_i(u_i - u)(u_i - u)^T \tag{14}$$

$$S_w = \frac{1}{N}\sum_{i=1}^{C}\sum_{j=1}^{N_i}(z_j^i - u_i)(z_j^i - u_i)^T \tag{15}$$

Moreover, S_{ij} describes the similarity of pixels i and j, and η is a coefficient to balance the trade-off between the training discriminant power and contextual constraints. The constraints function $J_2(w)$ gives a high penalty when the weights of related pixels differ too much. Due to the symmetry of S_{ij} in general case, the contextual constraints $J_2(w)$ on weight image can be formulated using the matrix operations further as follows:

$$\begin{aligned} J_2(w) &= \frac{1}{2}\sum\nolimits_{ij}(w_i - w_j)^2 S_{ij} \\ &= w^T L^w w \end{aligned} \tag{16}$$

where $L^w = D - S$ is the Laplacian matrix, and D is the diagonal matrix where $D_{ii} = \sum_j S_{ij}$. Thus, the objective of CCLDA can be formulated as:

$$J = \frac{w^T S_b w}{(w^T S_w w + \eta w^T L^w w)} \tag{17}$$

The optimal projection w can be obtained by solving the following generalized eigenvalue problem:

$$S_b w = \lambda (S_w + \eta L^w) w \tag{18}$$

And in our paper, the similarity matrix of weights is determined as follows:

$$S_{ij} = \begin{cases} e^{\|f_i - f_j\|^2/\sigma^2}, & if\ i\ and\ j\ are\ neighbors \\ 0 \quad , & otherwise \end{cases} \tag{19}$$

where f_i and f_j are the feature vectors extracted at position i and position j, respectively to describe the texture and spatial relationship between position i and position j. It should be noted the definition of weight similarity S is not limited. Different definitions are possible according to different problems.

To take into account the spatial relation of pixels in images, we proposed an improved method based on the CSLDA, which incorporates the contextual information based linear discriminant analysis into client specific linear discriminant analysis.

2.3 Locality Preserving Projection

In the Laplacianfaces approach [47], the face images are transformed into image vectors, and image vectors usually lead to a high-dimensional image vector space. Due to the large size of the image vector space and the relatively small number of training samples, Laplacianfaces approach first projects the image vectors into a PCA subspace to avoid the singular problems, which can throw away the smallest principal components. And then LPP is conducted in the PCA subspace.

The linear transformation based on PCA can be denoted as follows:

$$X = W^T Z \tag{20}$$

where $X = [x_1, x_2, \ldots, x_N]$ denotes the coordinates of Z in the subspace spanned by the transformation matrix $W = [w_1, w_2, \ldots, w_n]$, n is the number of the transformation vectors.

Then we conduct the LPP method with the transformed sample vector $x_i (i = 1, 2, \ldots, N)$. LPP considers the manifold structure which is modeled by an adjacency graph and can find the optimal linear approximation to the Eigenfunctions of the Laplace Beltrator on the manifold.

For the transformed training set $X = [x_1, x_2, \ldots, x_N]$, a reasonable criterion for choosing a 'good' map is to minimize an objection function.

The objective function of LPP is as follows:

$$\min \sum_{ij} \left(a^T x_i - a^T x_j\right)^2 S_{ij} = \min\left(a^T XLX^T a\right) \tag{21}$$

where a is a transformation vector, the matrix S_{ij} is a weight matrix. A possible way of defining S_{ij} is as follows:

$$S_{ij} = \begin{cases} \exp\left(-\frac{\|x_i - x_j\|^2}{t}\right) & , x_i \text{ is among k nearest neighbor of } x_j \text{ or } x_j \text{ is among k nearest neighbor of } x_i \\ 0 & , \text{otherwise} \end{cases} \tag{22}$$

where $i, j = 1,2,\ldots,N$, t is a parameter that can be determined empirically.

The vector $\{a_i\}$ that minimize the objective function are reduced to the generalized eigenvalue problem as:

$$XLX^T a = \lambda\ XDX^T a \tag{23}$$

Where $X = [x_1,x_2,\ldots,x_N]$; D is a diagonal matrix, and its entries are column(or row, since S is symmetric) sum of S,$D_{ij} = \sum_j S_{ij}$.; L is the Laplacian matrix, $L = D\text{-}S$.

Let λ_i and a_i be the ith minimum eigenvalue and its corresponding eigenvector of $(XDX^T)^{-1}XLX^T$. Then let $T = [a_1, a_2, \ldots, a_m]$ and $D_b = diag[\lambda_1, \lambda_2,\ldots,\lambda_m]$.

However, feature extraction of client specific linear discriminant analysis (CSLDA) does not consider the local feature. To avoid the deficiency, we incorporate the local preserving projections (LPP) with client specific subspace.

So then we follow the function (1) projecting all the transformed training samples X into the subspace spanned by $U = TD_b^{-1/2}$. Finally we follow the description described in function (2–12) to get the client specific subspace.

2.4 The Training Procedure of MKL-DR

To overcome the restrains of existing dimensionality reduction, we adopt a general framework of dimensionality reduction for data in various feature representations via multiple kernel learning. The essence of MKL is to find an optimal way to linearly combine the given kernels, each of which is created based on a specific kind of data descriptor and fuses the descriptors in the domain of kernel matrices. The training procedure of MKL-DR is as follows [40]:

Suppose we adopt M kinds of kernel functions to characterize various visual feature of each sample in the form of base kernels $\{K_m\}_{m=1}^M$. It can be described as:

$$K^{(i)} = \begin{bmatrix} K_1(1,i) & \ldots & K_M(1,i) \\ .. & .. & .. \\ .. & .. & .. \\ K_1(N,i) & \ldots & K_{M(N,i)} \end{bmatrix}, \qquad i = 1,2\ldots,N \tag{24}$$

where $K_m(i,j) = k_m(z_i,\ z_j)$.

Moreover, the method specified by two affinity matrices W and W'. In our paper, for convenience, the two affinity matrices $W = [W_{ij}]$ and $W = [W'_{ij}]$ are defined as:

$$w_{ij} = \begin{cases} 1/N_i, & if\ \omega_i = \omega_i \\ 0, & otherwise \end{cases} \tag{25}$$

$$w'_{ij} = \frac{1}{N} \tag{26}$$

Then sample coefficient vectors A is optimized by solving the generalized eigenvalue problem:

$$S_W^\beta a = \lambda S_{W'}^\beta a \tag{27}$$

where

$$S_W^\beta = \sum_{i,j=1}^{N} w_{ij}(K^{(i)} - K^{(j)})\beta \quad \beta^T(K^{(i)} - K^{(j)})^T \tag{28}$$

$$S_{W'}^\beta = \sum_{i,j=1}^{N} w'_{ij}(K^{(i)} - K^{(j)})\beta \quad \beta^T(K^{(i)} - K^{(j)})^T \tag{29}$$

And kernel weight vector β is optimized by solving optimization problem (30) via semidefinite programming:

$$\min_\beta \quad \beta^T S_W^A \beta \tag{30}$$

subject to $\beta^T S_{W'}^A \beta = 1$ *and* $\beta \geq 0$

where

$$S_W^A = \sum_{i,j=1}^{N} w_{ij}(K^{(i)} - K^{(j)})AA^T(K^{(i)} - K^{(j)})^T \tag{31}$$

$$S_{W'}^A = \sum_{i,j=1}^{N} w'_{ij}(K^{(i)} - K^{(j)})AA^T(K^{(i)} - K^{(j)})^T \tag{32}$$

After accomplishing the training procedure of MKL-DR, we can get the sample coeffient vectors $A = [a_1\, a_2 \ldots a_P]$ and kernel weight vector β.Then we are ready to project the sample z_i into the learned space of lower dimension by

$$x_i = A^T K^{(i)} \beta, \quad i = 1, 2, \ldots, N \tag{33}$$

where $K^{(i)} = R^{N \times M}$ and $K^{(i)}(n, m) = k_m(z_n, z_i)$.

To address the effect of kernel techniques on face verification, we proposed a face verification algorithm based on the combination of multiple kernel learning and client specific subspace (CSMKL). Different from the above methods, we need the framework of MKL in advance to reduce the dimensionality.

So then we conduct the CSLDA with the transformed sample vector $x_i (i = 1,2,\ldots,N)$ in MKL subspace to get the projection subspace as described in 2.1.

3 Experimental Results and Analysis

In order to test the performance of the proposed algorithms in this paper, face verification experiments have been conducted on both the XM2VTS database according to Lausanne Protocol [43].

3.1 Data Preparation

The XM2VTS database [43] is a multi-model database consisting of video sequences of talking faces recorded for 295 subjects at 1 month intervals. The data has been recorded in 4 sessions with 2 shots taken per session. From each session two facial images have been extracted to create an experimental face database of size 55 × 51. Figure 1 shows examples of images in XM2VTS.

For the task of personal verification, a standard protocol for performance assessment has been defined. The so called Lausanne protocol splits randomly all subjects into a client and imposter groups. The client group contains 200 subjects,

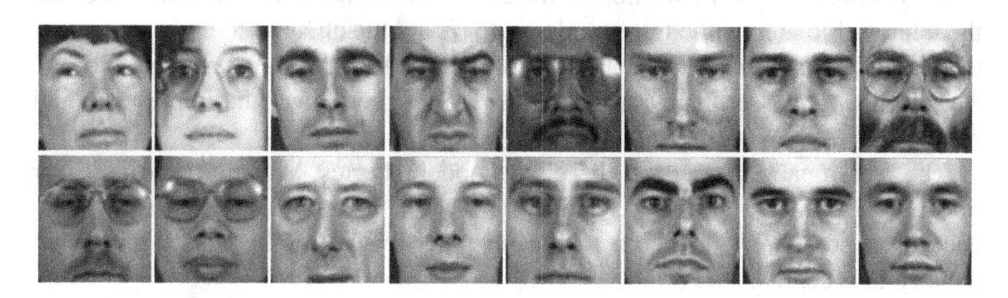

Fig. 1 Part of the images in XM2VTS

the imposter group includes 95 subjects, which is divided into 25 evaluation imposters and 70 test imposters. Within the protocol, the verification performance is measured using false acceptance and false rejection rates. In experiments, the database is divided into three sets: training set, evaluation set and test set. The training set is used to construct client models. The evaluation set is selected to produce client and imposter access scores which are used to find a threshold that determines if a person is accepted or rejected. The test set is selected to simulate real authentication tests.

In the experiments, two error measures of a verification system are the False Acceptance rate (FAR) and the False Rejection rate (FRR). False acceptance is the case where an imposter, claiming the identity of a client, is accepted. False rejection is the case where a client, claiming his true identity, is rejected. FAR and FRR are given by

$$\begin{aligned} FAR &= EI/I^{*}\,100 \\ FRR &= EC/C^{*}\,100 \end{aligned} \tag{34}$$

where EI is the number of imposter acceptances, I is the number of imposter claims, EC the number of client rejections, and C the number of client claims. For the test set, I is 112000(70 imposters*8 shots*200 clients) and C is 400(200 clients*2 shots).

Finally, a nearest neighbor classifier was employed for classification. The Euclidean distance measure is adopted. Both FAR and FRR are influenced by an acceptance threshold. Usually, FAR increases with the threshold increasing, while the FRR is on the contrary. Therefore, we select the threshold at the equal rate (EER) where both the false rejection and false acceptance rates are the same obtained on the evaluation set. Then the same threshold will be used on the test set to test the verification performance of the proposed method. Then the same threshold will be used on the test set to test the verification performance of the proposed method. Details of the protocol can be found in [41].

3.2 Classification

In the context, we used two models for classification in [41], which are based on the relationship among the test sample vector b and the client specific fisher face of the claimed identity a_i, the mean vector of the ith class and the mean vector of the i-th disguising classes. They are called the client model and the imposter model respectively.

The classification based on the client model: the test statistic to be used for making decision can simply be defined as $d_c = |a_i^t b - a_i^t u_i|$. If the test statistic d_c exceeds a predefined threshold t_c, the claim is rejected, otherwise the claimed identity is accepted.

The classification based on the imposter model: the test statistic to be used for making decision can simply be defined as $d_i = |a_i^t b - a_i^t u_\Omega|$. If the test statistic d_i exceeds a predefined threshold t_i, the claim is accepted; otherwise the claimed identity is rejected.

3.3 Experimental Results Obtained on XM2VTS

Tables 1, 2, 3 and 4 illustrate the verification performance of CSLDA and other dimensionality reduction techniques combined with client specific subspace on XM2VTS. All the experiments are based on the client model and imposter model respectively, which are referred to as On C and On I respectively. In the tables, TER equals to the sum of FAR and FRR, which refers the total false rate. In Tables 1 and 2, k refers to the parameter of the weight matrix S in CCLDA and LPP. And the parameter M refers to the kinds of kernel functions in Table 4.

Although there are many choices of kernel functions, we only use the basic functions to express the base kernels $\{K_m\}_{m=1}^M$ in MKL:

Polynomial kernel function: $K(z_i, z_j) = [a(z_i \cdot z_j) + b]^d$

RBF kernel function: $K(z_i, z_j) = \exp\left(-\|z_i - z_j\|^2 / 2\delta^2\right)$

Sigmoid kernel function: $K(z_i, z_j) = \tanh\ulcorner q(z_i \cdot z_j) + \Theta \lrcorner$

The values of the coefficient affect the performance of the face verification, hence we set the value of parameters as $a = 1e-4$, $b = 1$, $d = 2$, $q = 0.01$, $\Theta = -4$ and the width parameter of RBF is δ by experiments.

As shown in the tables, we can see some interesting points.

Table 1 Performance comparison between CSLDA and CSCCLDA method on XM2VTS

Methods	Evaluation			Testing		
	Model	FAR	FRR	FAR	FRR	TER
CSLDA	On C	3.0325	3	2.7848	3	5.7848
	On I	4.5	4.5	5.2804	4.5000	9.7804
CSCCLDA ($k = 2$)	On C	1.385	1.5	1.3062	1.5	2.8062
	On I	3.0275	3	3.5946	3	6.5946
CSCCLDA ($k = 3$)	On C	1.425	1.5	1.3187	1.5	2.8187
	On I	3.2025	3.25	3.7446	3.25	6.9946
CSCCLDA ($k = 4$)	On C	1.265	1.25	1.2205	1.25	2.4705
	On I	3.135	3.25	3.7277	3.25	6.9777
CSCCLDA ($k = 5$)	On C	1.325	1.25	1.3	1.25	2.55
	On I	3.215	3.25	3.7759	3.25	7.0259
CSCCLDA ($k = 6$)	On C	1.2675	1.25	1.2411	1.25	2.4911
	On I	3.25	3.25	3.8179	3.25	7.0679
CSCCLDA ($k = 7$)	On C	1.385	1.5	1.3518	1.5	2.8518
	On I	3	3	3.6054	3	6.6054

Table 2 Performance comparison between CSLDA and CSLPP method on XM2VTS

Methods	Evaluation			Testing		
	Model	FAR	FRR	FAR	FRR	TER
CSLDA	On C	3.0325	3	2.7848	3	5.7848
	On I	4.5	4.5	5.2804	4.5000	9.7804
CSLPP ($k = 2$)	On C	2.5750	2.5000	2.6420	2.5000	5.1420
	On I	1.6875	1.7500	1.9107	1.7500	3.6607
CSLPP ($k = 3$)	On C	1.7500	1.7500	1.7732	1.7500	2.3375
	On I	1.0550	1.000	1.3375	1.000	4.4018
CSLPP ($k = 4$)	On C	1.7500	1.7500	1.8500	1.7500	3.6000
	On I	1.1050	1.000	1.2214	1.000	2.2214
CSLPP ($k = 5$)	On C	1.7500	1.7500	1.7991	1.7500	3.5491
	On I	1.000	1.000	1.1812	1.000	2.1812

Table 3 Performance of CSKDA on XM2VTS

Methods CSKDA	Evaluation			Testing		
	Model	FAR	FRR	FAR	FRR	TER
Poly	On C	2.5	2.5	2.4232	2.5000	4.9232
	On I	1.75	1.75	1.9179	1.7500	3.6679
RBF	On C	1.2075	1.25	1.1187	1.25	2.3687
	On I	2.005	2	2.2982	2	4.2982
Sigmoid	On C	4.9775	5	4.85	5	9.85
	On I	4.5	4.5	4.8259	4.5	9.3259

Table 4 Performance of CSMKL method on XM2VTS

Methods	Evaluation			Testing		
	Model	FAR	FRR	FAR	FRR	TER
CSMKL $M = 1$(poly)	On C	1.5	1.5	1.5	1.5	3
	On I	2.25	2.25	1.7913	2.2	4.0412
CSMKL $M = 1$(rbf)	On C	0.9611	1	0.8662	1	1.8663
	On I	1.8611	1.75	1.3963	1.75	3.1463
CSMKL $M = 2$(poly + rbf)	On C	0.9333	1	0.8625	1	1.8625
	On I	1.8597	1.75	1.3888	1.75	3.1388
CSMKL $M = 2$(poly + sig)	On C	7.7486	7.75	7.8025	7.75	15.5525
	On I	4.1458	4.25	3.7213	4.25	7.9713
CSMKL $M = 2$(rbf + sig)	On C	0.9639	1	0.84	1	1.84
	On I	1.5875	1.5	1.1413	1.5	2.6413
CSMKL $M = 3$	On C	0.9667	1	0.8525	1	1.8525
	On I	1.6944	1.75	1.2238	1.75	2.9738

(1) In all experiments, the verification performances of different techniques of feature extraction vary considerably from each other.
(2) Compared with CSLDA, CSCCLDA, which considers the image contextual constraints, significantly improves the verification performance, especially on the client model. Besides, both of them can get better results on the client model than the imposter model. Moreover, the parameter of k has less influence on the results, which illustrates the better robustness of CSCCLDA.
(3) Table 2 shows the performance comparison of CSLPP and CSLDA for different value of the parameter k. We can see that the parameter k has some influence on the experimental results of CSLPP, but overall the results do not change significantly. Unlike the CSLDA and CSCCLDA, we can find that the performance based on the imposter model is better than the client model in CSLPP. Meanwhile, the CSLPP can fully keep the advantages of both client specific subspace and LPP, so our method can get better results than CSLDA.
(4) Comparing CSKDA with CSMKL, we can see that our method can conveniently deal with image data depicted by different descriptors using multiple kernels. So our method is more effective than CSKDA. Moreover, we can find that different combination of kernels have different effect on the results of the experiments. We also find that our method improves the performance of face verification significantly in some cases.
(5) Overall, the performance of CSKDA and CSMKL are slightly better than that of the other methods above. The advantage of CSKDA and CSMKL is that they can express the important information contained in the higher order relationships among the image pixels of a face pattern with kernel techniques.

Figures 2, 3, 4, 5 and 6 show the ROC curves of the CSLDA, CSCCLDA, CSLPP, CSKDA and the CSMKL on XMTVTS for both client model and imposter model. Comparing these figures, we can get some observations.

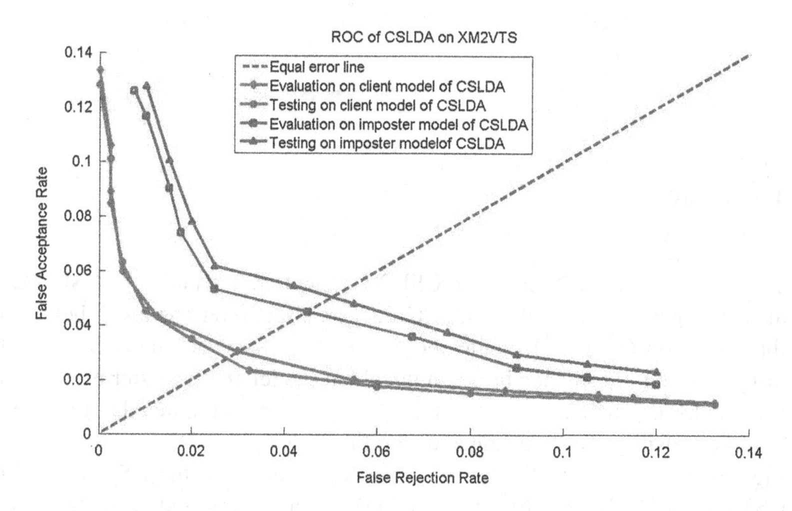

Fig. 2 ROC curves of CSLDA

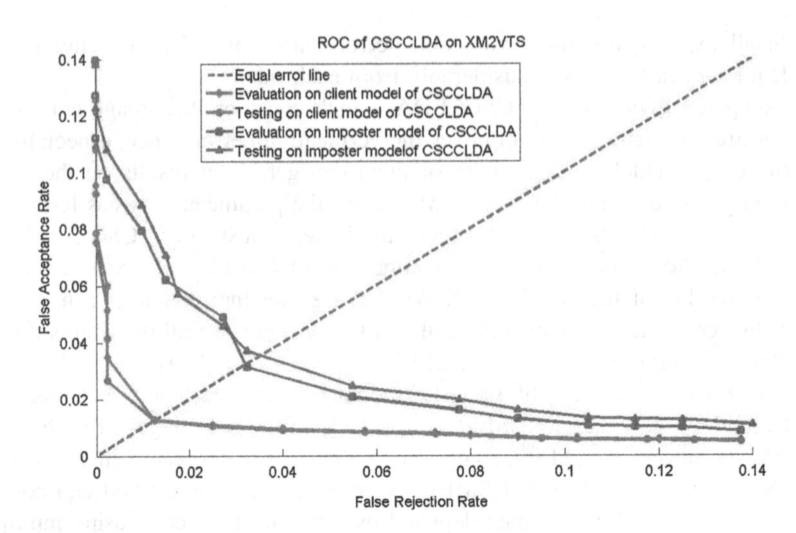

Fig. 3 ROC curves of CSCCLDA

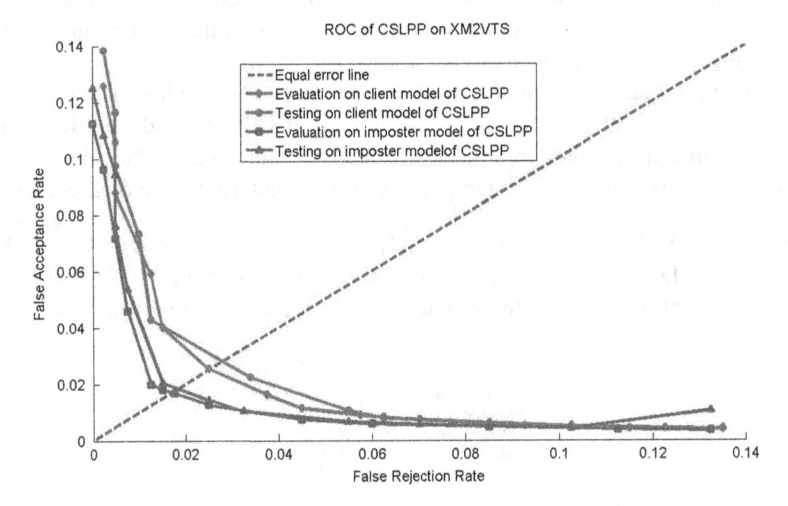

Fig. 4 ROC curves of CSLPP

(1) The error rate of CSCCLDA and CSLPP is much less than that of CSLDA, we also find that the error rate of CSCCLDA on client model is much better than that on imposter model, while the CSLPP is on the contrary. Besides, the difference in performance between the client model and imposter model based the CSLPP is much apparent than the CSLDA, which implies that the CSLPP is more sensitive to the changes of the threshold.
(2) Figures 5 and 6, respectively, show the ROC curves of the CSKDA and the CSMKL ($M = 3$) on XM2VTS for both client model and imposter model.

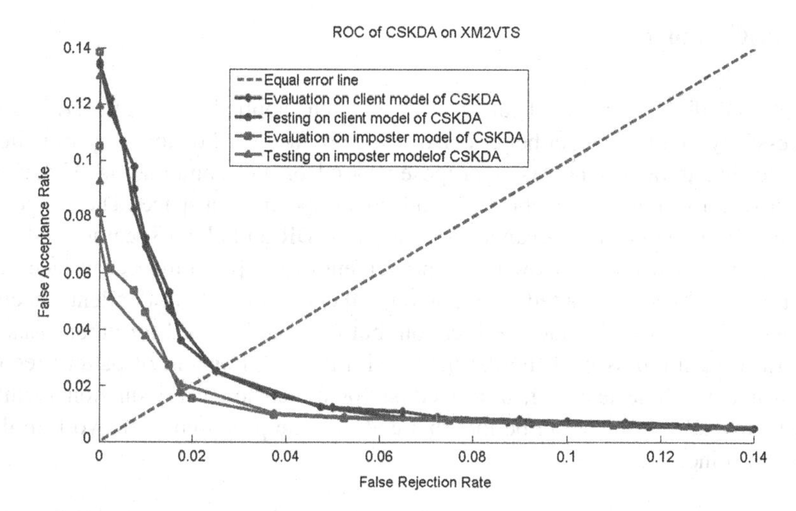

Fig. 5 ROC of CSKDA

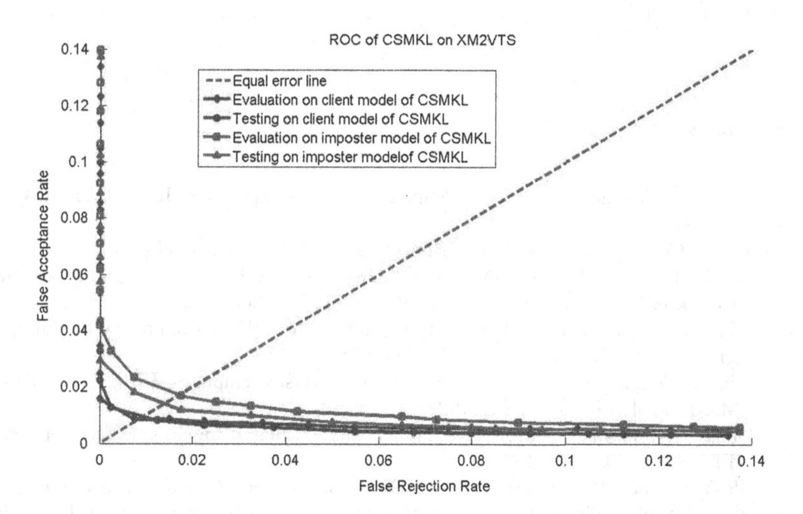

Fig. 6 ROC of CSMKL

Comparing the two figures, we can see the error rate of CSMKL is much less than that of CSKDA. We also find that the difference of performance between the client model and imposter model based on the CSMKL is much significant than the CSKDA, which implies that our method is more robust.

4 Conclusions

Client specific discriminant analysis has been applied to face verification successfully. In order to get better performance in face verification, a hybrid client specific discriminant analysis is proposed based on the combination of various dimensionality reduction techniques and client specific subspace. The proposed framework integrates the advantages of different DR and client specific methods. The experimental results show the improvement of the performance on face verification. It should be noted that not only the proposed hybrid client specific methods can be used for face verification, but they can be used for other tasks of pattern recognition as well. Furthermore, with the development of pattern recognition and machine learning, new method for dimensionality reduction method will be available, which can be incorporated into the proposed framework in this paper in principle.

Acknowledgments This work was supported in part by the following projects: 111 Project of Chinese Ministry of Education (Grant No. B12018), Key Grant Project of Chinese Ministry of Education (Grant No.: 311024), National Natural Science Foundation of P. R. China (Grant No.: 60973094, 61103128)

References

1. Ronsenfeld, A.: Image analysis and computer vision. Comput. Vis. Imag. Underst. **62**(1), 33–93 (1997)
2. Kanade, T.: Computer Recognition of Human Faces. Birkhauser, Basel (1977)
3. Brunelli, R., Pogio, T.: Face recognition through geometrical features. In: 1992 Proceedings of the European Conference on Computer Vision, pp. 792–800 (1992)
4. Craw, I., Tock, D., Bennett, A.: Finding face features. In: 1992 Proceedings of the European Conference on Computer Vision, pp 92–96 (1992)
5. Brunelli, R., Poggio, T.: Face recognition: features versus templates. IEEE Trans. Pattern Anal. Mach. Intell. **15**, 1042–1052 (1993)
6. Chellappa, R., LWilson, C., Sirohey, S.: Human and machine recognition of faces: a survey. Proc. IEEE **83**(5), 705–740 (1995)
7. Sun, S.-Y., Zhao, H.-T., Yang, H.-J.: Discriminant uncorrelated locality preserving projection. In: Proceedings of the. International Congress on Image and Signal Processing, pp. 1849–1852 (2010)
8. Ng, A.Y., Jordan, M.I., Weiss, Y.: On spectral clustering: analysis and an algorithm. In: Advances in NIPS 14, MIT Press, Cambridge, pp. 849–856 (2001)
9. Bitouk, D., Miller, M.I., Younes, L.: Clutter invariant ATR. IEEE Trans. Pattern Anal. Mach. Intell. **27**(5), 817–821 (2005)
10. Moon, J.K., Kim, K., Kim, Y.: Design of missile guidance law via variable structure control. J. Guidance Control Dyn. **24**, 659–663 (2001)
11. Sun, S.-G., Park, H.W.: Invariant feature extraction based on radial and distance function for automatic target recognition. In: Proceedings of ICIP 2002, vol. III, pp. 345–348 (2002)
12. Devijver, P.A., Kittler, J.: Pattern Recognition: A Statistical Approach. Prentice-Hall, Englewood Cliffs (1982)

13. Sirovich, L., Kirby, M.: Low-dimensitional procedure for the characterization on human faces. J. Opt. Soc. Am. A **4**(3), 519–524 (1987)
14. Kirby, M., Sirovich, L.: Application of the Karhunen-Loeve procedure for the characterisation of human faces. IEEE Trans. Pattern Anal. Mach. Intell. **12**(1), 103–108 (1990)
15. Turk, M., Pentland, A.: Eigenface for recognition. J. Cogn. Neurosci. **3**(1), 70–86 (1991)
16. Pentland, A., Moghaddam, B., Starner, T.: View-based and modular eigenspaces for face recognition. In: Proceedings of CVPR'94, pp. 84–91
17. Romdhani, S.: Face recognition using principal component analysis. MSc Thesis, University of Glasgow. http://www.elec.gla.ac.uk/romdhani/pcadoc/pcadoctoc.htm (1997)
18. Moghaddam, B., Pentland, A.: Probabilistic visual learning for object detection. In: Proceedings.of the 1995 International Conference on Computer Vision, pp 786–793 (1995)
19. Yang, J., Zhang, D.: Two-Dimensional PCA: a new approach to appearance-based face representation and recognition. IEEE Trans. Pattern Anal. Mach. Intell. **26**(1), 131–137 (2004)
20. Moon, H., Philips, P.J.: Analysis of PCA-based face recognition algorithms. In Proceeding of the FG'98, pp. 205–210 (1998)
21. Fukunaga, K.: Introduction to Statistical Pattern Recognition. Academic Press, New York (1990)
22. Belhumeur, P., Hespanha, J.P., Kriegman, D.J.: Eigenfaces vs. fisherfaces: recognition using class specific linear projection. IEEE Trans. Pattern Anal. Mach. Intell. **19**(7), 711–720 (1997)
23. Swets, D.L., Weng, J.: Discriminant analysis and eigenspace partition tree for face and object recognition from views. In: Proceeding of the FG'96, pp. 192–197 (1996)
24. Etemad, K., Chellappa, R.: Face recognition using discriminant eigenvectors. In Proceedings of the ICASSP'96, pp. 2148–2151 (1996)
25. Etemad, K., Chellappa, R.: Discriminant analysis for recognition of human face images. J. Opt. Soc. Am. A **14**(8), 1724–1733 (1997)
26. Liu, C., Wechsler, H.: Face recognition using evolutionary pursuit. In: Proceedings of ECCV'98, vol. II, pp. 596–612 (1998)
27. Zhao, W., Chellappa, R., Krishnaswamy, A.: Discriminant analysis of principal components for face recognition. In: Proceedings of the Third IEEE International Conference on Automatic Face and Gesture Recognition, pp. 336–341 (1998)
28. Li, Y.P., Kittler, J., Matas, J.: Effective implementation of linear discriminant analysis for face recognition and verification. In: the Proceeding of CAIP'99, pp.234–242 (1999)
29. Lei, Z., Li, S.Z.: Contextual constraints based linear discriminant analysis. Pattern Recogn. Lett. **32**, 626–632 (2011)
30. Huang, R., Pavlovic, V., Metaxas, D.: A hybrid face recognition method using Markov random fields. In: ICPR, Cambridge, pp. 157–160 (2004)
31. Dass, S.C., Jain. A.K.: Markov face models. In: Proceedings of ICCV 2001, pp. 112–116 (2001)
32. Dass, S.C., Jain, A.K., Lu, X.: Face detection and synthesis using Markov random field models. In: Proceedings of the 16th International Conference on Pattern Recognition, pp. 201–204 (2002)
33. Wang, L., Zhang, Y., Feng, J.: On the Euclidean distance of images. IEEE Trans. Pattern Anal. Mach. Intell. **27**(8), 1334–1339 (2005)
34. Tenenbaum, J.B., De Silva, V., Langford, J.C.: A global geometric framework for nonlinear dimensionality reduction. Science **290**(5500), 2319–2323 (2000)
35. Roweis, S.T, Saul, L.K.: An introduction to locally linear embedding. http://www.cs.nyu.edu/~roweis/lle/papers/lleintro.pdf (2000)
36. Belkin, M., Niyogi, P.: Laplacian eigenmaps and spectral techniques for embedding and clustering. In: Neural Information Processing Systems, Mit Press, Vancouver, pp. 585–591 (2001)
37. He, X.Y., Niyogi, P.: Locality Preserving Projections. Neural Information Processing System. Mit Press, Vancouver (2003)

38. Chen J.-F., Li, B., Yuan, B.-Z.: Face recognition using direct LPP algorithm. In: Proceedings of ICSP 2008, pp. 1457–1460 (2008)
39. Yang, M.-H., Ahuja, N., Kriegman, D.: Face recognition using Kernel eigenfaces. In: Proceedings of the 2000 IEEE International Conference on Image Processing, pp. 37–40 (2000)
40. Lin, Y–.Y., Liu, T.-L.: Multiple Kernel learning for dimensionality reduction. IEEE Trans. Pattern Anal. Mach. Intell. **33**(6), 1147–1160 (2011)
41. Kittler, J.: Face Authentication Using Client Specific Fisherfaces. Centerfor Vision Speech and Signal Processing. University of Surrey, Guildford (2001)
42. Wu, X.-J., Kittler, J., Yang, J.-Y., Kieron, M., Wang, S.-T., Lu, J.-P.: On dimensionality reduction for client specific discriminant analysis with application to face verification. Lect. Notes Comput. Sci. **3338**, 305–312 (2004)
43. Kieron, M., Kittler, J., Luettin, J.: XM2VTSDB: the extended M2VTS database. In: Proceedings of the Audio and Video Based Biometric Person Authentication Conference (AVBPA99) (1999)
44. Li, Y.P.: LDA and its application to face identification. PhD dissertation, Surrey, Center for Vision Speech and Signal Processing, University of Surrey (2000)
45. Wu, X.-J.: Client specific Kernel discriminant analysis algorithm for face verification. In: Proceedings of International Conference on Neural Networks and Brain, Beijing (2005)
46. Yu, W.-W., Teng, X.-L., Liu, C.-Q.: Discriminant locality preserving projections: a new method to face representation and recognition. In Proceedings of the 2005 International Workshop on VS-PETS, pp. 201–207 (2005)
47. He, X.-F., Yan, S.-C., Hu, Y., Niyogi, P., Zhang, H.-J.: Face recognition using laplacianfaces. IEEE Trans. Pattern Anal. Mach. Intell. **27**(3), 328–340 (2005)

Printed in the United States
By Bookmasters